Scientific Authority

Twentieth-Century

America

Scientific Authority

& Twentieth-Century

America

EDITED BY

Ronald G. Walters

Johns Hopkins University Press

Baltimore and London

Copyright © 1997 The Johns Hopkins University Press
All rights reserved. Published 1997
Printed in the United States of America on acid-free paper

Johns Hopkins Paperbacks edition, 2001
9 8 7 6 5 4 3 2 1

The Johns Hopkins University Press
2715 North Charles Street
Baltimore, Maryland 21218-4363
www.press.jhu.edu

Library of Congress Cataloging-in-Publication Data will be
found at the end of this book.

A catalog record for this book is available from the British Library.

ISBN 0-8018-5390-7 (pbk.)

Contents

Scientific Authority

Twentieth-Century

America

Uncertainty, Science, and Reform in Twentieth-Century America

Ronald G. Walters

I N 1922, George Ellery Hale, director of the Mount Wilson Observatory, wrote an article for *Scribner's Magazine* describing the headquarters for the National Academy of Sciences and the National Research Council, then under construction in Washington, D.C.[1]

Hale used the occasion to extol the virtues of science in a brief narrative of its history that extended from "the mists of antiquity," to the "keen vision of the Greeks," into a long dark night of "mediaeval obscurity," through its reemergence in a heroic struggle against superstition during the Renaissance. Thereafter, it was a story of nearly five hundred years of unbroken progress to the present moment, in spite of the persistence of a few contemporary counterparts of "the inquisitors who burned Giordano Bruno and imprisoned Galileo." These, he added in words foreshadowing the Scopes trial in Tennessee three years later, were men and women "who regard science as the enemy of their particular creeds, and would penalize the teacher of evolution and the student of the origin of man." As far as Hale was concerned, such people were marginal, ignorant, and easily dismissed. "But whatever be one's viewpoint," he declared, "he must be blind indeed if he fails to recognize the services of science to civilization." Science produced such miracles as the airplane, solved the mysteries of geology and evolution, increased agricultural yields, lengthened life, and reduced the price and increased the abundance of consumer goods. Even so, according to Hale, its "work for humanity has only just begun."[2]

Hale may have underestimated the extent, variety, and sophistication of critics of science, but the most striking thing about his view is how commonplace and widely shared it was at the time. He and, very likely, his educated, middle-class readers and fellow scientists were confident that science was cumulative and ever progressing. "Its greatest aim and object is the discovery of the truth, which it pursues without fear of embarrassing consequences." It was authoritative and beyond politics, greed, and selfishness. It was a hallmark of civilization.

By the end of the twentieth century, Hale's confident tone was harder to find. Compare his bland dismissal of critics of science with the alarm sounded in 1994 by the executive director of the National Center for Science Education. "Wake up scientists!" she wrote. "The attack against objectivity, critical thought, and rationality is rising, and its entrenchment in the halls of higher education means future leaders will be poorly equipped with the intellectual skills to handle this ever more complex world. . . . The real bottom line is our future as a civilization."[3] Publication of several major defenses of science indicate that others had already awakened and shared the executive director's dismay. One reviewer noted in 1993 that the four books he discussed "differ in almost every respect except that their authors all believe that somebody is misrepresenting the contemporary state of science to the public."[4] He added, in a rhetorical question that marked the distance between his intellectual world and Hale's, "given the increasing specialization and mutual ignorance of twentieth-century scientists, who would be so bold as to claim access to the knowledge needed for solving the world's problems?"[5] The central tenets of Hale's faith were up for debate by the 1990s, including his belief that science was the key to human progress. In the gloomy assessment of two latter-day defenders of science, the enemy was no longer a few cranks and religious fundamentalists on the fringes of American society. What they called "higher superstition" was prevailing even within the universities that had produced much of the best twentieth-century science.[6]

The essays in this collection reflect that "higher superstition" by taking a critical view of claims by those purporting to speak for science, especially claims of being able to reshape the world in a morally neutral, infallible, and inevitable fashion. The authors demonstrate the manner in which science consists of human choices, blocks some alternatives as it opens others, privileges some voices at the expense of others, serves particular kinds of interests, and exerts its authority in a variety of ways, some quite subtle. In these respects, these writers belong to a line of scholarship stretching back over forty years to Thomas S. Kuhn's *The Structure of Scientific Revolutions*, which, among other things, challenged the notion that science proceeds through a progressive, linear accumulation of newer, better data and theories.[7]

In another crucial respect, however, these essays point to important alternatives in thinking about science and its role in social change. It is far too easy to see the choice as between accepting Hale's belief that science is a quest for truth whose findings should be authoritative and accepting a completely pluralistic, relativistic view that there are no absolute truths and that one person, or group's, truth is as good as any other. Three of the chapters—those by David Hollinger, Dorothy Ross, and James Kloppenberg—explicitly address those positions and suggest ways of understanding science in history, culture, and society without either blindly granting it ultimate authority or dismissing it entirely as a source of knowledge and guidance. Although less explicit in searching for a general framework within which to locate science, the remaining chapters weigh the practices and costs of scientific authority not with Luddite malice but rather with the purpose of improving those practices and lowering the social and cultural costs. This is an effort to foster, not foreclose, debate over the place of science in shaping a democratic society.

The relationship between science and social reform is a tangled one and relatively recent in origin. Indeed, the notion of *reform* as an ongoing and positive process, and of *reformer* as a career, clearly emerged in America only in the first half of the nineteenth century.[8] Modern science as we know it is still more recent. Even so, as early as the 1830s, some reformers evoked "science" as a justification for particular causes and a guide for social change. Such rhetoric, however, appeared most commonly in movements to create new and better kinds of institutions—schools, asylums, and prisons, for instance—as well as among a small number of utopians and in health and other reforms aimed at improving humankind through understanding physiological "laws." In a few instances, "science" actually served as a weapon against reformers' goals. Beginning in the late 1830s, scientists of the so-called American School of Anthropology undermined antislavery and calls for women's rights by citing measurements of human skulls to "prove" that the races had been created separately and that men were superior in intelligence to women. George Fitzhugh similarly marshaled statistics and evoked the "science" of sociology to defend slavery against abolitionist attacks. Prior to the Civil War, however, science usually took third place to religious and political rhetoric—notably, the languages of evangelical Protestantism, republicanism, individualism, and artisan radicalism—in debates over what America ought to be.

That changed after the Civil War. A small but significant sign of the transformation occurred within the Women's Christian Temperance Union (WCTU), an organization with intellectual and organizational roots in antebellum traditions of religious and women's activism. Among its more successful endeavors in the last two decades of the nineteenth century was a campaign to promote scientific temperance instruction in public schools. Although the science was debatable, the results, by 1901, were considerable, and the effort to cloak what had been primarily a moral crusade with the mantle of science was telling.[9] The WCTU, however, was far from alone in using science to further what it saw as reform. The last decades of the nineteenth century and the first two of the twentieth—the Progressive period—produced a striking burst of confidence in the ability of science, including social science, to describe, prescribe, manage, and improve the state of life. That the faith persists to the present day in many quarters should not obscure the fact that it is a historical creation, not much more than a century old in American culture.

To some extent, the power of science to explain, guide, and predict social change gained as the force of alternative ideologies waned. Older doctrines derived from evangelical Protestantism, republicanism, and individualism lost energy and relevance in the more complex, industrial, urban, conflicted, interdependent, and ethnically heterogeneous world of the late nineteenth century. Socialism made inroads but failed to achieve mass support. Science alone seemed to promise certainty and a realm of truth beyond debate.

Equally important in enhancing the prestige of science, it and its offspring, technology, appeared to be doing things of enormous value in late nineteenth-century America. Prior to the Civil War, American science lent some credence to Alexis de Tocqueville's dour judgment "that in few of the civilized nations of our time have the higher sciences made less progress than in the United States."[10] Such a comment would have been untenable by the early twentieth century, after the rise of university- and government-sponsored research, the development of the sciences, engineering, and social sciences as distinct professions, and the victory of the American Medical Association over alternative practitioners. The growth, somewhat later, of the corporate research laboratories that figure into the essays by John Stilgoe and Roland Marchand and Michael Smith only added to America's extraordinary scientific and technological capital.

Considerable care is in order, nonetheless, before attributing the prestige of science in twentieth-century America simply to its ability to produce socially beneficial results. The three chapters in this volume that treat one of the genuine triumphs of nineteenth-century medical research—the germ theory of disease—tell cautionary tales. Regina Morantz-Sanchez and Allan Brandt show how this theory helped inhibit alternative ways of organizing and imagining health care oriented toward public health and holistic medicine. JoAnne Brown demonstrates how images, language, and metaphors derived from germ theory served commercial and political interests quite remote from science. Equally chastening, however, is Brandt's observation that "most important infectious diseases were declining even before elucidation of the germ theory."

In that instance, as well as with the fitness movement of the 1970s and 1980s, medical science was able to capitalize on improvements in health that originated elsewhere.

In other cases, crediting science with something beneficial helped obscure a more complex set of causes and interests. Dr. Hale's speech provides an example when he claimed the airplane to be a result of pure science rather than of military advantage and corporate capitalism, both of which probably played more significant roles in its development. Other instances appear in the chapter by Marchand and Smith, which demonstrates how industry used its version of science to promote a particular vision of the future that had more to do with corporate politics than with pure research. Such appropriations of science, however, may simply reinforce the point: they only worked because, for several generations of Americans, beginning with those born before the Civil War, it was plausible to believe that science had produced miracles within their lifetimes—because, indeed, it had.

The specific connections between science and reform in the twentieth century, however, have been shifting, complex, and ambiguous. Science, for example, provides predictions and a vocabulary for talking about the future, ranging from utopian visions (including the commercial ones recounted in the Marchand and Smith chapter) to calls to action to avoid environmental disaster, energy depletion, and overpopulation. Proponents sometimes claim science to be the ideal court of last appeal; in practice, it frequently is an arena of contention, with scientific experts either speaking on different sides of policy issues or declining to take a stand. It provides data for policy makers to interpret and to translate into action, occasionally in extraordinarily significant ways, as when the Supreme Court used testimony from social scientists to arrive at its historic 1954 desegregation decision in Brown versus Board of Education of Topeka. The authority of science legitimates particular positions and certain groups' powers, and it provides languages and metaphors that shape and limit debate. Further complicating matters, science is often available to impede as well as to foster reform; in the late twentieth century, science and reform agendas sometimes compete for resources in federal, state, university, and corporate budgets.

By the middle of the twentieth century, much of the power of science to affect social change was firmly imbedded in institutions.[11] A trail of federally sponsored research stretches from expeditions like Lewis and Clark's (1804–6); various forms of mapping and surveying; the authorization of the National Academy of Sciences (1863); the Hatch Act of 1887; the creation of the agricultural research apparatus, which looms large in Stilgoe's chapter; other Progressive Era research in such areas as forestry; the formation of the National Science Foundation in 1950; and a dramatic increase in support of scientific research after World War II. By the 1990s, an extraordinary tangle of government agencies and laboratories, independent laboratories, universities, industrial laboratories, and private foundations had arisen. By 1983, these various institutions were conducting "almost half of the Western World's research and development" and spending "more money on science than Japan and the industrialized nations of Europe combined."[12]

However powerful individual scientific institutions may be in particular fields, the scattering of research into diverse and sometimes competing locations, combined with increasing specialization within disciplines, has important consequences for the social role of science. On the one hand, it makes for "big science," in the sense of expensive and extensive, with a highly visible presence in American life. On the other hand, it makes science a convenient abstraction rather than a unified entity in whose name one can speak. The authority of science, therefore, frequently depends less on what it can actually say or do than on the particular interests with whom it is allied and the respect politicians, university officials, corporate executives, investors, and consumers have for it. That respect, in turn, rests in large measure on the claim of a disinterested commitment to absolute knowledge that Hale made in his 1922 magazine article.

At the turn of the century, the notion of entrusting social problems to disinterested experts, scientists among them, had great appeal to middle-class Americans dismayed at the corruption and turmoil of democratic politics in which the "best men" no longer ruled. The undemocratic nature of deferring to scientific authority was part of its attractiveness and remains so today. Evoking scientific impartiality as a counter to mere politics worked within government as well as with the public. President Jimmy Carter's science adviser claimed to have "found it advantageous to be viewed primarily as a professional rather than a political appointee, particularly in my dealings with Congress, industry, universities, and professional societies." He further noted that "the credibility of my advice [to the cabinet] was enhanced by the apolitical and impartial image" of his office.[13]

Once established, the cultural authority of science moved from individual scientists and their achievements to subtle forms of power. As several of the following chapters demonstrate, this power was mainly the ability to define certain kinds of knowledge as legitimate, the ability to exclude others kinds of knowledge from legitimacy, and the ability to affect language, imagery, and metaphor. Testimony to the rhetorical strength of science is the persuasiveness of terms derived from it even when they are not especially appropriate: how much, for example, is a "social experiment" (to use a common reform figure of speech) really like a scientific one?[14] Also revealing is the manner in which — as Brown and Marchand and Smith show — Americans have incorporated the image of science, with its promise of future progress, with images of the past, notably the frontier. When Vannevar Bush titled his famous 1945 report *Science: The Endless Frontier,* he was far from unique. Even the erudite and thoughtful physicist J. Robert Oppenheimer could not resist the frontier metaphor as he tried to explain to undergraduates the "strange destiny" American leadership in science imposed upon the nation. Nor did another physicist, Frederick Seitz, see anything incongruous in calling his 1994 autobiography *On the Frontier: My Life in Physics.*[15] When an analogy has such power among scientists, it is little wonder that politicians, corporations, and advertising agencies find it similarly appealing. Whatever its ironies and limitations, however, the cultural power of science, as embedded in language and metaphor, helps structure how Americans think and talk about their society. Through its appropriation by advertisers and the media, moreover, science's prestige links notions of progress to the development of a consumer economy in which producing and purchasing material goods is the essence of pioneering.

None of this is to say that the authority of science ever went unchallenged. Because of such genuine accomplishments as the germ theory and widespread political and economic support after World War II, its late twentieth-century defenders sometimes risk overestimating (as Dr. Hale surely did) its cultural authority. They sometimes imagine a golden age when Americans deferred to men — and as Morantz-Sanchez reminds us, they were mostly men — of science. From such a perspective, criticism of science in recent decades is bewildering in its sophistication and all the more distressing because much of it comes from the academies that produce scientists.

Some late twentieth-century defenders of science focus upon a small number of particular villains whose influence they see as particularly culpable in deflating the notion of science as neutral, progressive, and inevitable.[16] A short list of the usual suspects would include Thomas Kuhn, the philosophers Paul Feyerabend and Richard Rorty, Michel Foucault, Sandra Harding, Carolyn Merchant, Evelyn Fox Keller, and assorted other academics, postmodernists, and feminists. Such writers have had considerable impact, and the essays by Hollinger, Ross, Morantz-Sanchez, and Kloppenberg acknowledge and analyze their influence. But as those essays also make clear, intellectual critiques of science come from a variety of perspectives and are themselves part of a much larger late twentieth-century project of seeking out the hidden operations of power, questioning once seemingly fixed and "essential" categories such as race, gender, and ethnicity, asserting the importance of interpretation, denying universal claims to truth, and rejecting simple, linear narratives like the one of inevitable scientific progress. Far from being singled out for special attention, science stands as one social and cultural construction to be demystified among many such constructions, including the very disciplines (such as history) from which this line of criticism comes.[17]

To lay responsibility for skepticism about science on the academic left, moreover, ignores serious, critical analyses of scientific research. Journalist and author David Dickson, for example, seems blissfully uninfluenced by postmodernism, feminism, and Foucault as he dissects "the new politics of science."[18] To blame the academic left likewise dismisses from view other critics, ranging from conservatives nostalgic for an imagined preindustrial past, antivivesectionists and animal rights activists, religious groups who repudiate conventional medicine, and New Age gurus. In political terms, the strength and persistence of antiscience views among conservative groups in American society is especially significant. Members of the religious right, for example, reject the notion that abortion is a medical issue in any respect, as it was frequently defined in the nineteenth century, and continue to wage war against the teaching of evolution in public schools, as their forebears did in Dr. Hale's day.

Finally, to attack present-day critics of science as misguided and cranky radicals does more than violate the historical record: it obscures problems within science itself and the degree to which it invites hard scrutiny, particularly when applied to social issues. On that score, the sources of frustration among intellectuals and the public alike are several. The historical record contains reminders that what seem to be progressive uses of science from one perspective look reactionary in hindsight. Between 1907 and 1928 scientific warnings about a "feebleminded menace" to the United States resulted in a spate of mandatory sterilization laws and the sterilization of more than eight thousand allegedly unfit Americans, a practice that was as dubious scientifically and socially as it was constitutionally. In another notorious instance, a 1972 examination of a forty-year Public Health Service study of men with tertiary syphilis in Tuskegee, Alabama, revealed an ugly story of callousness and racism in the name of science.[19] While those episodes might seem safely in the past by the 1990s, their ghosts still haunt American science and social science in recurring debates about alleged linkages between intelligence, criminality, and race.

From the 1960s onward, scientists themselves did little to enhance their image as spokespersons on crucial issues. Intense specialization, dependence on federal and corporate funding, and commitment to the ethos of disinterestedness, led many to avoid speaking out on public issues or, when they did, to speak with so much caution and nuance that their words carried little impact. Such reticence left the field open for popularizers and scientists with marginal credentials, who were anything but disinterested.[20] That — combined with attacks on the indirect cost recovery universities collect for scientific research, disagreements among scientists on global warming and other environmental and health issues, and well-publicized allegations of scientific fraud — played to cynical suspicions that scientists were either out of touch with reality or, like lawyers, bought by the highest bidder. After decades of strong fiscal support for research, the end of the twentieth century produced ample evidence that many Americans felt that science, as represented by such things as a high-energy supercollider or the space program, was simply too expensive. Given the realities of taxpayer discontent and of social problems that science did not appear to be addressing, science faced increasing scrutiny from the government, corporations, and universities.

The following essays do little to refurbish the image of science, which, in any event, is still doing quite nicely in spite of intellectual and fiscal challenges — and which remains one of the things America does best. Instead, they set out a series of questions about science and about its linkages to reform and social change. Beneath the case studies and diverse critiques, however, is the suggestion that the alternatives are not, as they frequently seem to be framed in late twentieth-century America, between either straightforward acceptance of science as a progressive path to truth or a world in which interpretation is all. The choice — to put it differently — is not between claims of universal truths (whether from science or elsewhere) and a hopelessly relativistic world of distinct, virtually hermetically sealed groups, each with its own vision of reality.

Although differing in focus and emphasis, as well as on a number of specific issues, the essays in this collection constitute a call for paying attention to alternative sources of knowledge and authority, for understanding the historic moment, and above all, for openness of discourse. In that respect, the volume will seem to some to have a "liberal" cast, to use a term that became virtually a political smear word in the 1980s. The charge is true if liberalism means acceptance of difference and pluralism. It is not true to the extent that twentieth-century liberalism uncritically tied its hopes to science and expertise, as it did at many points from the Progressive Era through the "experiments" of the New Deal and the Great Society. Nor is it true to the degree that liberalism in this century endorsed the power of centralized authority, especially in the form of a strong state (whose official agricultural science Stilgoe is eloquent in questioning). From these essays it is possible to imagine a world in which science is not privileged as a source of universal, unchanging truth but as self-consciously located within its historic and social moment, tested against experience (including against itself), and engaged in continual, open debate. This would not be a world without values or truths, as conservative defenders of science might charge, or a reversal of a half millennium of progress, as Dr. Hale's narrative might have it, but it would be a world that includes humility about our knowledge and an awareness that our values and truths have their historic origins and that power lurks beneath claims to expertise and within the language we use to describe the world. It would be faithful to one of the traditional goals of science itself: continuous, open inquiry. And that, in the curious way things work, is what appeals to scientific authority in the name of social progress have often tried to foreclose.

PART I

Voices of Authority

THE TWO chapters in part 1 deal with science as a source of authority in political and social life as well as with alternatives and challenges to its claims of universal truth and comprehensiveness. Dorothy Ross demonstrates the power of scientific models, tools, and language in shaping American social science, just as later chapters show their power in other aspects of American life. Ross also makes the intriguing suggestion that, in the American context, the alternative to what she calls "scientism"—locating change within history rather than in natural law—has had an affinity for radicalism, at least in the 1930s and 1960s. If so, the implication of her argument is that scientific modes of thought and language are, at best, limited instruments of social change in a political culture that claims the authority of nature for established principle.

For David Hollinger, science has moved over the course of the twentieth century from being a source of authority authorizing or impeding reform to being an object of criticism and reform from other groups and other perspectives. In common with Ross, he proposes an alternative that is respectful of science as a source of knowledge while rejecting extreme science-based claims to possess universal, unchanging truth. His alternative is a "postethnic" society that acknowledges the existence of multiple, overlapping identities (including professional ones) while encouraging open discourse and debate to explore commonalities and differences across group boundaries. Ross's alternative to scientism similarly accepts, and even endorses, diversity of viewpoints while insisting on the importance of thinking historically. These are themes on which chapters in the following parts play variations and to which James Kloppenberg returns in the final chapter.

CHAPTER ONE

How Wide the Circle of the "We"?

American Intellectuals and the Problem
of the Ethnos since World War II

David A. Hollinger

exual Behavior in the Human Male; Sexual Behavior in the Human Female: the titles imply an inquiry specieswide in scope.[1] And, indeed, the authors were professional zoologists. But the Kinsey Reports, as these studies came to be called, were based on interviews with a highly particular zoological sample: men and women in mid-twentieth-century North America, overwhelmingly the cultural products of the United States. The deflation of the universalist pretensions of these studies began almost immediately upon publication of the first of the two volumes in 1948.[2] It was the destiny of the Kinsey Reports to become artifacts in an animated and enduring discourse not about humankind but about a particular society and its culture. Librarians were obliged to catalogue them in the science section; at home, however, individuals found good reasons to shelve these fascinating tomes next to David Riesman's *The Lonely Crowd,* Gunnar Myrdal's *American Dilemma,* and Henry Nash Smith's *Virgin Land.* What started out as zoology ended up as American studies.

Alfred Kinsey and his staff shared with a multitude of their American contemporaries a destiny against which intellectuals of our own time struggle conspicuously. This destiny is to have confused the local with the universal: to have made claims about or claims on behalf of all humankind for which the salient referent was later said to be but a fragment of that elusive whole. In the name of the rights and needs of the entire species, the United States tried to advance through war and diplomacy what we are now told were the historically particular interests of the North Atlantic bourgeoisie. In the name of the epistemological unity of all humankind, philosophers vindicated a scientific practice generated and sustained by what may have been the peculiar preoccupations of European and American white males. In the name of a mystical "humanity," the prophetic Hoosier Wendell Willkie proclaimed "one world," in which farmers near Kiev deserved our sympathy and respect because they were just like farmers near Kokomo, Indiana.[3] In the name of "the family of man," Edward Steichen and the Museum of Modern Art exhibited a series of photographs displaying what we now recognize as a sensibility common to male American liberal intellectuals of the period.[4] In the name of an essential "human nature," Freudians and behaviorists offered prescriptions designed for the entire population of the globe. In the name of specieswide fraternity, advocates of Esperanto sought through a distinctly European synthetic language to mitigate the divisive effects of linguistic diversity throughout the world. In the name of an ostensibly universal capacity for spiritual experience, religious ecumenists sought to neutralize sectarian conflict through the claim that "we all believe in the same God," which turned out, of course, to be the God of liberal Protestantism.

To resist the extravagant universalism discernible in the scientific, social scientific, humanistic, religious, and political discourse of the generation represented by the Kinsey Reports is to belong to the sprawling cohort that I refer to in this chapter as *we*. We situation-conscious intellectuals realize that Kinsey's generation was far from the first to conflate the local with the universal, but the legacy of this mid-twentieth-century American generation presents a special challenge for us. Not only have our Kinseys and Willkies and Hans Reichenbachs been close at hand; their universalism was often directed against certain particularisms we still take to be evil. These particularisms include white supremacy and other forms of racism and ethnic prejudice; they include, also, nationalist chauvinism, religious bigotry and obscurantism, a host of provincial taboos, and the several authoritarian warrants for belief associated with the names of Lysenko and Goebbels. It was to break down such enclosures that American intellectuals of the midcentury decades claimed to look to the species as a whole and reasserted Enlightenment notions of knowledge and rights. Hence, one might expect, logically, that resistance to the universalist enthusiasms of the midcentury generation would take the form of a yet more insistent and rigorous universalism, according to which the covertly particularist biases of our Kinseys and Willkies and Reichenbachs would be eliminated. If that generation confused the local with the universal, let the next generation, with its greater cultural self-awareness, look beyond the local to the genuinely universal.

Some have tried to do just that in their science and scholarship. They have represented more critically the fields of experience in which their objects of study can be found while continuing to strive for insights that can apply to enlarged domains and can be justified by standards recognized in many cultures. Some have also sought to refine the old universalism in relation to "human rights" abroad and to "civil rights" at home. American protests against South Africa's apartheid system were often couched in universalist terms. But alongside these efforts to serve universalist ideals there have appeared during the past several decades a multitude of initiatives of a very different sort.[5]

Species-centered discourse itself has become suspect. If the universalism of the World War II era served to deracinate and to efface the varieties of humankind through the use of too parochial a construction of our common humanity, and if this universalism served further to mask a cultural imperialism by which the NATO powers spread throughout the world their own peculiar standards for truth, justice, and spiritual perfection, then universalism itself, we are told, is too dangerous an ideal. Let us emphasize instead the integrity of all the varieties of humankind, let us encourage every culture to find its own distinctive voice, let us enable the repressed "other" to command our attention through new canons and curricula and, thereby, to destroy the cultural hierarchy created by Western prejudice. So, we beat the drums for alterity and wonder whether the defense of Salman Rushdie's freedom of speech is not another bourgeois conceit, the salient functions of which are to prevent Muslims from worshiping in peace and to enable Western intellectuals to feel superior to the still-benighted East. But the critique of universalism has not been limited to persons moved by the claims of alterity. Defenders of the mainstream social democratic traditions of the West have mounted some of the most widely discussed critiques of the old universalism. Alasdair MacIntyre, Robert Bellah, and Richard Rorty tell us, in their contrasting idioms, that Enlightenment universalism itself was a noble dream for which we are now too sophisticated.[6]

The science we believe and practice derives its warrant, we are sometimes told, not from its presumed capacity for verification across the lines of all the world's cultures but from the authority of the distinctive social entities Thomas S. Kuhn and his followers have helped us to recognize: sharply bounded professional communities characterized by rigorous procedures for the acculturation of their members.[7] The obligations we owe to one another and the rights we claim for ourselves derive not from our common membership in a species; our new moral philosophers tell us that these obligations and rights derive from the ordinance of the traditions of our singular tribe. The covert enthnocentrism we now discern within the species-centered discourse of the World War II generation stands not as invitation to develop a yet more rigorous universalism but rather as proof of the impossibility of escaping ethnocentrism. The very impulse to produce "metanarratives" and to develop "totalizing" perspectives is, in itself, to be overcome. Our mission, apparently, is not to purge the old universalism of its corruptions but to renounce it as fatally flawed and to perfect, instead, the local and the particular, to live within the confines of the unique civic, moral, and epistemic communities into which we are born, to devote ourselves to our ethnos.

This radically historicist understanding of obligations, rights, and knowledge is what Rorty now defends as "ethnocentrism," a label perhaps not shrewdly chosen but one that has the merit of dramatizing for us the ideological distance our new cognoscenti have traveled from the world of Kinsey, Willkie, and Reichenbach.[8] That distance consists chiefly in the growing acceptance since the 1960s and 1970s of our own historicity. By historicity, I mean simply the contingent, temporally and socially situated character of our beliefs and values, of our institutions and practices.[9] When we accept our own historical particularity, we shy away from essentialist constructions of human nature, from transcendentalist arguments about it, and from timeless rules for justifying claims about it. We eschew the Archimedean perspective, and we instead inquire outward from our experience. We practice immanent, rather than transcendental, critique. We confront the insight that truths and rights and obligations become available to us — and thus, it is apparently easy to conclude, come into existence for us — through the operation of historically contingent communities of human beings. The science that was once a source of authority, capable of justifying or discrediting social reform, has become for some an appropriate object of reform authorized by other, social-political authorities.[10]

Antecedents of this historicism are easy to find: in nineteenth-century German hermeneutics, for example, in the turn-of-the-century American pragmatists, and in the cultural relativism associated with Boasian anthropology. But one work of the post-1945 era that commandingly represents this new historicist movement is Kuhn's classic of 1962, The Structure of Scientific Revolutions.[11] Although this book is addressed to science, and specifically to the most mathematically developed and technical of the sciences, its apparent demonstration of the dependence of scientific truth on the historically specific practice of distinctively organized human communities gave unprecedented credibility to the historicist perspective as applied elsewhere. If even the most persuasively verified claims of astronomers and physicists were in large part cultural products, what could be said about the claims of social scientists and humanistic scholars, of moral philosophers, metaphysicians, political theorists, and others whose ability to rise above opinion and prejudice has always been precarious and contested?

There were still distinctions to be made, of course: one could posit a spectrum of claims running from the relative universality of warrant to the relative locality of warrant, with the periodic table of elements at one end and the value of a good suntan at the other; but in the wake of the transformation for which Kuhn's name has become an emblem, this whole spectrum shifted toward the local, and it became more plausible to view the Bill of Rights as just another tribal code than as a manifestation in one polity of claims advanced on behalf of all humankind. This rather anthropological way of looking at our own culture was resoundingly proclaimed by the era's most eloquent and influential anthropologist, Clifford Geertz, who urged that we see "ourselves amongst others, as a local example of the forms human life has locally taken, a case among cases, a world among worlds."[12] Eric R. Wolf may have been right to remind us that "the peoples who have asserted a privileged relation with history and the peoples to whom history has been denied encounter a common destiny," but can our knowledge of that "common destiny" be anything other than "local"?[13] Even Geertz's localism came to seem relatively solicitous of global perspectives when several self-reflexive anthropologists displayed yet more vividly and theorized yet more earnestly the situated character of their own ethnographic practice, rendering themselves all the more distant from the epistemic arrogance associated with European colonialism.[14] By the 1980s, even the great Kantian, John Rawls had sidled cautiously toward the historicists, leaving leadership of the resistance in the hands of dwindling cadres of Straussians and Jesuits.[15] And if, indeed, the old-time Diltheyans fell forward on their faces as opposition to the reign of hermeneutics virtually ceased, they were run over from behind by the poststructuralists.[16]

Whatever else poststructuralism has contributed, it has helped to constitute the more general phenomenon known as postmodernism. Some formulations of the touted transition from modernism to postmodernism run roughly parallel to the transition from species-centered discourse to ethnos-centered discourse. *Modernism* and *postmodernism* are, of course, highly contested terms, conveying a number of different meanings in the contemporary debate, but postmodernism is frequently invoked to condemn "the Enlightenment project" and to confer attention and dignity on the local, the fragmentary, and the particular. Amid the welter of constructions and counter-constructions of the modern and the postmodern, the species is very often to the modern what the ethnos is to the postmodern.[17] This ethnicization of all discourse through the decentering strategies of postmodernism is thus a culmination of the process by which the term *ethnic* has lost its connotations of marginality (originally, *ethnic* meant, according to the *Oxford English Dictionary*, "outsider," "pagan," or "gentile") and has come to stand for situatedness within virtually any bounded community, regardless of its relation to other communities.

Meanwhile, a chorus of both academic and popular political voices identified ethnoracial communities as vital sites for the formation, articulation, and sustenance of cultural values, social identities, and political power. Texts once celebrated for their apparent ability to speak to and for all people came to be valued as representatives of distinct social constituencies defined largely by the author's ethnoracial identity and gender. "Diversity" replaced "unity" in the slogans of those concerned to promote mutual respect and equality among the varieties of humankind found within the United States. The once-popular notion that there might be an American character or even an American culture was widely discredited as a nationalist equivalent of diversity-denying universalism. The ethnos came to be defined against not only the species but the closest thing to its national equivalent, the American nation. The figure of the melting pot, encumbered with assimilationist connotations, lost favor to the salad bowl, the mosaic, and the garden of plants each with its own autochthonous roots. The United States came increasingly to be represented as a complex patchwork of distinctive communities, sometimes said to be nation-like, each one a unique product of unique historical forces and circumstances. An ideological and constitutional tradition protecting the rights of individuals was asked to reform itself in order to protect instead the rights of groups. The name of Herder was rarely invoked by those who promoted or appreciated the new ethnic particularism, but Herder more than Marx or Mill or Jefferson or Condorcet is the relevant ideological precursor.

It is possible to exaggerate the intensity, scope, and singularity of this cluster of historicist and particularist enthusiasms of the last three decades and to perform the same exaggerations on behalf of the panorama of universalist initiatives of the previous era. But the differences within each of these two clusters ought to be obvious enough to render the juxtapositions all the more valuable as indicators of an event in recent intellectual history: the diminution of species-centered discourse and the enlargement of ethnos-centered discourse. This event transcends the United States, which can be construed as one of many national sites for its production. Considered as part of the intellectual history of the United States since World War II, moreover, this event is but a fraction of that history. Yet signs of its occurrence can be discovered in enough spaces within the discourse of American intellectuals to render the event worthy of historical and critical assessment. Rather than speculate about what, in the dynamics of the postwar period, might be the several causes of this change in the center of intellectual gravity for many American intellectuals, I want simply to call attention to this transformation, to address the importance of issues in affiliation that attend upon it, and to sketch a critical response to it.

THE IMPORTANCE of affiliation is implied by the new prominence of the notion of community in discussions of truths, meanings, rights, and goods. In the post-Kuhnian era, argumentation about knowledge is, much more than it used to be, argumentation about epistemic communities. Stanley Fish, treating communities of readers as analogues to Kuhn's communities of scientists, proposes that the meaning of texts is a matter decided by a profession of literary critics, who also decide what is and is not a "text" to begin with.[18] Michael Walzer organized his "radically particularist" "spheres of justice" around the issue of membership in a "community"; and in working out his analysis, Walzer explicitly eschews any effort to look outside the Platonic cave of his own tribe.[19] Rorty, addressing both epistemic and moral communities, argues that communal solidarity can perform the services for which people once turned naively to the ideal of objectivity. In an exchange with Geertz a few years ago, Rorty sketched a pluralist vision of the world as an expanse of private, exclusive clubs, interacting with as much civility as they could but each defined, animated, and sustained by a vivid sense of the difference between "we" and "they."[20] When it comes to justifying a truth claim, whom do we have to persuade? Whose testimony do we need to take into account? Why, the members of our own club, of our own epistemic community. When it comes to justifying rights and obligations, whom do we have to persuade? Whose opinions and sensibilities do we need to take into account? The members of our own club, our own moral community. Rorty predicts that "wet liberals" would cavil at his suggestion that "the exclusivity of the private club might be a crucial feature of an ideal world order," but such wet liberals continue to be the victims of lingering Enlightenment universalist illusions outgrown by dry liberals like Rorty himself.

Along with the new prominence of community comes a new centrality for the old question of membership. *The less one's raw humanity is said to count for anything, the more important become one's affiliations.* The more epistemic and moral authority is ascribed to historically particular communities, the more it matters just who is and is not one of "us." The more detached truth and goodness become from the testimony and tastes of any population outside our own tribe or club, the more is at stake when that tribe or club defines itself in relation to other human beings. How wide the circle of the "we"?

This may be *the* great question in an age of antiuniversalist historicism.[21] The frequency with which it is asked with a fresh sense of urgency, even by such sophisticated discussants as Bernard Williams, helps to distinguish our own historical moment.[22] Many old, nonhistoricist questions translate into the terms of this newly prominent question. Consider, for example, two old chestnuts especially favored by intellectuals of the 1930s and 1940s, subjectivity versus objectivity and relativism versus absolutism. Subjectivity was most often depicted as an individual, psychological state and objectivity as access to the domain of the external world. Can history be objective, can ethics be objective, is science really objective, they asked, while contrasting the values and preconceptions of the individual subject to an objective reality.

What has happened since then is a collectivization of the subjective, dissolving the binary opposition of subjective-objective into a population of minds grouped by the cultures and subcultures whose epistemes or paradigms or codes largely create these minds as thinking subjects. To be sure, subjectivity was sometimes defined in terms of deference to tribal norms rather than in terms of individual idiosyncrasy, but the trend is unmistakable. Peter Novick catches the difference in his recent study of the objectivity question among professional historians, noting the transition in relativism from Carl Becker's "Everyman His Own Historian" in 1932 to "Every Group Its Own Historian" since the 1960s.[23] In the meantime, within the swelling ranks of the historicists, those counted as defenders of "objectivity" (a term now usually in quotation marks) turn out to be defenders of large-scale consensus, as with Kuhn, who recently formulated his disagreement with Rorty by purposefully conflating objectivity with Rorty's ideal of solidarity: "Like solidarity, objectivity extends only over the world of the tribe, but what it extends over is no less firm and real for that," Kuhn insists. But Kuhn's tribe is implicitly large—he says he can imagine no life without the "single character," for which he takes both *objectivity* and *solidarity* to be names—while Rorty's club is small.[24]

Or, so it sometimes seems. Several of Rorty's recent writings deserve special attention here because in them are visible the challenges and dilemmas of the newly prominent problem of the ethnos.[25] This problem can be cast in Rortian terms: "solidarity with whom?" In the mid-1980s, when he first began to call by the name of ethnocentrism his insistence that we need not orient ourselves to the species, Rorty stressed how limited was the interchange he envisioned between the various clans that divide people from one another.[26] In subsequent years, however, he has moved toward a more inclusive notion of the ethnos.

In 1989, in *Contingency, Irony, and Solidarity*, Rorty allowed that it would be fine to try to "extend our sense of 'we' to people whom we have previously thought of as 'they' "; and on the final page of the book, he decisively extends the circle of the "we": what we must build, he declares, is "an ever larger and more variegated *ethnos*," for a crucial feature of our ethnos is, after all, its traditional "distrust" of "ethnocentrism." So, "we" build outward, from "our" starting point, which is a community of "heirs to the historical contingencies which have created more and more cosmopolitan, more and more democratic political institutions."[27]

Hence Rorty wants to mobilize all the liberal, cosmopolitan instincts that have been directed against ethnocentrism by John Dewey, Franz Boas, Ruth Benedict, and their kind, but so fearful is Rorty of universalist claims that he takes pains to characterize this venerable antiethnocentrism as a tribal peculiarity of ours and thus as part of the ethnos about which Rorty proposes to be ethnocentric. Rorty assures us that the circle can never be expected to embrace all humankind, but he manages to put in place a particularist cover for chastened gestures toward universalism. Rorty is often accused of authorizing a smug and parochial abdication of responsibility toward our fellow humans; in responding to this accusation, Rorty reverts repeatedly to this same device, the characterization of chastened universalist gestures he wants to support as merely our own particular particularism. The liberal "ideals of procedural justice and human equality" are probably "the best hope of the species," he says, even if these ideals are "parochial, recent, eccentric . . . local, and culture-bound."[28]

In one of his most vivid discussions of the problem, Rorty takes up the hypothetical case of "a child found wandering in the woods, the remnant of a slaughtered nation whose temples have been razed and whose books have been burned." Such a lost child is said to have, on Rorty's principles, no claim to "human dignity" because the moral community that sustained this notion in his or her case no longer exists. But, says Rorty, if the lost child is fortunate enough to be found by "us," by representatives of our ethnos, we will take the child into our community, there "to be reclothed with dignity." Why? Because this child is indeed human, and our ethnos carries the universalist aspirations of the Judeo-Christian tradition, "gratefully invoked," writes Rorty candidly, "by free-loading atheists like myself."[29]

What I find most revealing about Rorty's chosen example of the child without a community is that Rorty's chastened universalism with a particularist cover can express itself the most spontaneously when its object is not part of a living, viable community. One might look, instead, at the more challenging problem represented by the Masai women.[30] If these women are but breeding stock and, when barren of sons, are treated by their warrior masters as inferior to cattle, who are we to criticize? It is part of Masai culture, after all. And we probably should not even talk about it, as such talk might flatter Western prejudice and might lead us to forget how much violence and injustice are suffered by women in the United States and Western Europe. But if the Masai peoples eventually die off amid the economic, political, and ecological transformation of East Africa, we could, on ethnocentric principles, rescue the last surviving Masai woman as she crawls starving across the Ogaden. The price of her emancipation would then be the death of her culture, which we are restrained from countenancing by our principled antiuniversalism and our healthy suspicion of Western imperialism.

Some feminists would be less timid about the matter and might even dare to be judgmental. Whether or not a program has been devised that might actually help the Masai women—or the victims of genital mutilation in several other societies—the value of a critical perspective on the plight of women worldwide is affirmed by some feminists. Hence, some varieties of feminism sustain an element of species-centered discourse, a willingness to address the interests of women everywhere.[31] Many of the feminists who take an interest in the Masai women and in the victims of genital mutilation are also historicists, sensitive to the particularity of the cultures with which women as well as men must deal and reluctant to dismiss as false consciousness the insistence by some Saudi and Sudanese women that Western feminists have no authority to instruct them on their rights, needs, and duties. Feminism, like moral philosophy, philosophy of science, and a host of other intellectual enterprises, confronts in an age of historicism the question of the boundaries of the ethnos.[32]

Even in Rorty's view, it is membership in our species that endows with human dignity the Masai women, the lost child, or anyone else, but only because this biological and species-ist basis for dignity happens to be recognized by our particular Western ethnos. Does this make our culture superior to others? Only in our eyes, it would seem. But when our eyes are directed at "our own," we do, according to Rorty, have special responsibilities. It is in discussing these responsibilities in contemporary America that Rorty makes his strongest, yet most pragmatic, witness against the old, unchastened universalism:

Consider . . . the attitude of contemporary American liberals to the unending hopelessness and misery of the lives of young blacks in American cities. Do we say that these people must be helped because they are our fellow human beings? We may, but it is much more persuasive, morally as well as politically, to describe them as our fellow *Americans* — to insist that it is outrageous that an *American* should live without hope . . . our sense of solidarity is strongest when those with whom solidarity is expressed are thought of as "one of us," where "us" means something smaller and more local than the human race.[33]

Here Rorty draws the ethnocentric circle to correspond with American citizenship. He does so for a specific and admirable purpose: he hopes that some humans might be more effectively inspired to act to diminish the suffering of others. Yet the suffering that Rorty wants to diminish has been created and perpetuated by the feeling among many white Americans that blacks are not really part of "us" anyway. Distinctions between Americans of different racial and ethnic groups yield formidable we-they dynamics of their own. Indeed, such dynamics gave rise in the first place to the concept of ethnocentrism, which Rorty now wishes to appropriate for two benign purposes: the flagging of the recognition of our own historical particularity and the inspiring, Rorty hopes, of a measure of generosity greater than that previously called forth by appeals to the common humanity of all sufferers.

That such appeals to a grand humanity have failed to prevent a multitude of cruelties is easy to demonstrate, but Rorty seems less concerned to measure the practical efficacy of these old universalist appeals (can we be so sure that they have counted for so little in the history of human generosity?) than to warn against the Kantian rationalism with which he persistently associates them.[34] Kant made morality a matter of rational deduction from general principles, Rorty complains, and made "feelings of pity and benevolence . . . seem dubious, second-rate motives for not being cruel." But these motives are altogether praiseworthy, according to Rorty, who spends much of *Contingency, Irony, and Solidarity* arguing that novelists, ethnographers, essayists, and other "non-philosophical" intellectuals can do a lot more for the cause of human solidarity than can the Kantian thinkers who try to identify a "human essence" and to establish a "rational obligation" in relation to it.

Rorty's ethnocentrism is directed more sharply against rationalist and essentialist constructions of human nature than against any appeals to the solidarity of a historically real population. The extremity of Rorty's ethnocentrism turns out to be a disagreement with other philosophers over the terms on which human solidarity should be affirmed. As soon as the Kantians are disposed of, Rorty's vision of human solidarity takes on what most American intellectuals of the last half-century would recognize immediately as a decidedly antiethnocentric cast: this solidarity, says Rorty, should be understood as "the ability to think of people wildly different from ourselves as included in the range of 'us.'"[35] The circle of the we thus embraces diversity; it is not a unifor-mitarian construct, predicating equality or sameness.

Even Rorty, then, has come around to insisting that full recognition of the historically particular character of our discourses should not be taken as a license for what Geertz lampoons as "parochialism without tears,"[36] the giving up on the venerable cosmopolitan project of looking sympathetically beyond one's immediate surroundings. Rorty's effort to detach the concept of ethno-centrism from its conventional meaning has succeeded chiefly as a sophisti-cated irony, challenging people with lingering universalist tendencies to rec-ognize the historical particularity of their own antiethnocentrism. But even if ethnocentrism has served Rorty poorly, the dead ends and contradictions displayed by his forthright, admirably risk taking deployment of this concept against the conceits of his colleagues are highly instructive. These dead ends and contradictions can help to inspire a postparticularist, or as I prefer to say, postethnic, disposition toward issues of affiliation in a variety of contexts.[37] This disposition is not an answer to the problem of the ethnocentric circle; rather, it is a willingness to engage that problem while remaining suspicious of the will to enclose.

THE PERSPECTIVE I want to call *postethnic* differs decisively from the many "posts" of our time that serve to repudiate, rather than to build on and criti-cally refine, the contributions of what is taken to be the previous moment in a discourse.[38] A postethnic perspective—as I try to develop it here—takes the claims of the ethnos more seriously than do recent, widely discussed critiques of epistemic particularism by Ernest Gellner and of civic particularism by Arthur M. Schlesinger Jr.[39] It builds on several insights congenial to the basic historicism Rorty shares with Geertz, Kuhn, Walzer, and a number of other creative participants in the repudiation of universalism.

Historicists should be quick to acknowledge that the communities provid-ing moral and cognitive standards come into being under a great variety of circumstances, are perpetuated for many distinctive ends, and are driven by very different distributions of power. No sooner do we ask, How wide the circle of the we? than we ought to ask, What identifies the we? and How deep the structure of power within it? and How is the authority to set its boundaries distributed? These questions are invoked even in the hopeful verse of the old progressive poet remembered for having demanded the inclusion of "the man with the hoe," Edwin Markham:

He drew a circle that shut me out—
Heretic, rebel, a thing to flout.
But Love and I had the wit to win:
We drew a circle that took him in.[40]

Our communities are various in their structure and function: not all entail the same mix of voluntary and coerced affiliation, nor do all require the same measure of internal agreement, the same sorts of demands on the individual member, or the same degree of clarity in external boundaries.[41] Deciding who is in and who is out is a rather different matter depending on whether the salient "we" is understood to consist of Chicanos or of Texans, of particle physicists or of scientists, of Presbyterians or of Christians, of Skull and Bones or of what Rorty calls "the conversation of the West." The recognition of the vitality of the ethnos need not—as metaphorical references to tribes and clubs can imply—demand the dividing up of the world into an expanse of internally homogeneous and analogically structured units. We are better able to resist such reification[42] if we attend carefully to the overlapping character of many of these communities, to their internal diversity, and to their differing size, scope, and purpose.[43] What it means to situate oneself—a person or text being addressed—is not always as easy as our now ritualistic calls for the situating of everything are sometimes taken to imply. In an era of global creolization[44] and of deterritorialized communities, just what conditions are local are less than obvious.

Beginning with exactly these historicist sensitivities, a postethnic perspective goes on to recognize that most individuals are involved in many of these communities simultaneously and that the carrying out of any person's own life project entails a shifting division of labor between the several "we's" of which the individual is a part. How much weight is assigned to the fact that one is Pennsylvania Dutch or Navaho, relative to the weight assigned to the fact that one is also an American, a molecular biologist, a woman, and a Baptist? It is this process of consciously and critically locating oneself amid these layers of "we's" that most distinguishes the postethnic from the unreconstructed universalist. The latter wants to build life projects outside of, rather than through, particular communities. The willingness of the postethnic to treat ethnic identity as a problem rather than a given also helps to distinguish the postethnic from the unreconstructed ethnocentrist, for whom ethnic identity is a more straightforward proposition, a matter of affirming and developing something frankly parochial.[45]

A postethnic perspective also tries to remain alert to features of any given ethnos common to one or more other ethnoi that see each other as opposed.

When communities are construed as localities to which a norm, an aspiration, or a condition may be "historically particular," it remains true that these localities can be sites for the display of traits and conditions also found in other localities. Kiev and Kokomo have less in common than Willkie supposed, but that need not mean we have no basis for theorizing about, and acting on, needs and interests found to be shared by some inhabitants of these two sites. The hero reappearing behind "a thousand faces" may have been less singular than Joseph Campbell thought,[46] but inquiry into continuities across cultures still promises to help identify features of human life that are prominently and extensively developed within populations open to historical or contemporary scrutiny. Our choice of projects in philosophical anthropology need not be limited to a search for cultural universals, on the one hand, and on the other, a celebration of sheer difference.

If the observations in the three previous paragraphs have a truistic ring, they can serve all the better to remind us of the strength of the resources on which one might draw while scrutinizing the enclosures licensed by our new respect for boundaries and while specifying contexts in which even species-centered discourse might be critically and cautiously renewed. One such context is surely the geophysical health of the planet. Analyses of the globe's ecological trajectories render the notion of a specieswide interest strikingly authentic and help to neutralize the old suspicion that such universalist notions must always be a mask for some smaller, more particular interest. Disagreements between rich and poor nations about how to balance development with environmental protection complicate but do not invalidate the insight expressed in the environmentalist slogan One earth, one humanity, one destiny. Humankind as a nonmystical category can survive our recognition of its varieties.[47] It can survive as an object of critically grounded, scientific study. It can also continue to serve—imperfectly, to be sure—as a principle relevant to the operation of epistemic communities: it can serve these communities as a brake on the creation of a complacently ethnic *Wissenschaft*.

Professional communities in the natural sciences have proven able to warrant their major claims successfully in many locales and to acculturate persons born and raised in a multitude of tribes and clubs throughout the world. Yet many of us in the humanities and in the humanistically oriented social sciences are now so proficient in detecting parochialism and prejudice in even the natural sciences, and so conscious of the status of these sciences as historically specific, discursive practices, that we risk losing sight of the relative success of these sciences in making their ideas work, experimentally, and in incorporating within their "we" a great variety of men—and more recently, women—with diverse ethnoracial and national-cultural affiliations. All perspectives may be partial, but some are more partial than others. A postethnic orientation toward epistemic communities recognizes, rather than trivializes, the remarkable successes of the modern natural sciences but takes these successes as problems for theory and as challenges for politics rather than as proof that traditional and existing scientific communities have pushed the epistemic "we" as far as it can go. There may be a good bit more pushing to be done, and the resulting knowledge may serve more of the species than has been served by the knowledge we already have.

When we conclude that an idea works scientifically, in the context of what specific aims do we reach this conclusion? And when we observe that professional communities have managed to acculturate a demographically diverse population of men and women, just what elements in that human diversity are silenced in the interests of perpetuating the success of existing research programs? And how might these elements be given greater voice in the hope of generating and vindicating new and alternative research programs? These questions are being formulated and pursued by Ian Hacking, Mary Hesse, Evelyn Fox Keller, Nancy Cartwright, Peter Galison, and a host of other historians and philosophers of science who recognize the situated character of inquiry but who resist facile constructions of the relevant situations in terms of gross and self-enclosed social-cognitive units.[48] The work of these scholars sustains the hope that the knowledge sought by science can still—in this age of historicism—be construed as ideally public, subject to verification by anyone comparably equipped, trained, and positioned.[49] Hence, a postethnic perspective on epistemic communities seeks to test more systematically the limits of the epistemic "we" and to stretch its circle as wide as the capacities of nature and its knowers will allow.

An inclination toward such stretching is also what a postethnic perspective brings to moral communities, where its starting point might be Walzer's patently universalist attribution to all human beings of certain minimal rights, the rights "not to be robbed of life or liberty."[50] The point of underscoring this element of species consciousness in the fiercely particularist Walzer is not to reconvene a debate over whether there are universal values—historicists can quickly grant that Walzer's ascription of basic rights to all persons is an expression of Walzer's ethnos, and Kantians can worry about how this ascription might be justified and elaborated—but to call attention to the potential for practical dialogue across the lines that separate "us" from "them." Trying to reason with members of other tribes, trying to persuade them to recognize common interests, and even trying to convince them that they might be better off by adopting our ways are particulars of an ideal that is easily mocked. But a rejuvenated version of this old ideal is now being advanced with force and conviction by Jürgen Habermas, Jeffrey Stout, and others whose project of building a community through intersubjective reason would seem aimed exactly at the goal that even Rorty now acknowledges as his own: the expansion of "our" democratic-egalitarian ethnos through immanent critique.[51]

> I can try to exploit whatever vocabularies and patterns of reasoning we
> have in common, using them toward the end of bringing him around.
> If I am successful, though, it will probably be because he has observed
> me long enough to be taken, despite himself, by the way I live, to find
> his loves gradually shifting, irrespective of argument, in the direction of
> mine, to become, through unconscious modeling more than argumenta-
> tive persuasion, a less vicious person. Then, and only then, are arguments
> likely to make a difference. He might, of course, be too far gone, too bitter
> or hateful or narrow-minded to reach. It is, moreover, quite conceivable
> that a whole society could suffer moral atrophy in this way.[52]

Whatever the ideal dynamics of the expansion of the moral "we," the hypothetical character of Stout's conversation with the fascist thug can remind us of the need to address nonhypothetical, historically realistic possibilities for community building in relation to democratic-egalitarian values. Among these possibilities is the distinctive sociopolitical entity in which universalist and particularist impulses often reach the compromises and operating arrangements by which daily civic life actually takes place in a field of power: the nation.

The durability of nationalism in the face of a multitude of internationalizing forces may suggest how deep is the human need for an ethnos of accessible proportions and how portentous are the affiliations people choose or are coerced into accepting. Not all nationalisms are alike. Some are more uncompromisingly ethnocentric, in the older, pre-Rortian sense of the term. The United States is one of several nations possessing a transethnic national consciousness, but it is the most historically conspicuous. "America is still a radically unfinished society," Walzer reminds us in tones reminiscent of Randolph Bourne's characterization of a dynamic, transnational America welcoming and transforming many varieties of humankind.[53] But Bourne spoke against the torrent of nativist, Anglo-Saxonism that eventually curtailed immigration in the 1920s; while Walzer speaks in the atmosphere of an increasingly widespread acceptance of cultural diversity as a national virtue. The United States is unusual in the extent to which its ideological spokespersons accept, and defend, the nation's negotiated, contingent character within a broad canopy of Enlightenment-derived, universalist abstractions. This universalist component in the traditional ideology of the American nation[54] is a resource potentially of value to anyone troubled by the ease with which ethnicity turns to hatred in settings as varied as Baku, Bensonhurst, and Bosnia.[55]

This universalist self-image has recently become less fraudulent in the wake of what is sometimes called the de-WASPing of America. Unprecedented demographic diversity—marked most dramatically by the numbers and varieties of Asian and Latin-American immigrants and their offspring—has further diminished the privileged connection between American nationality and Anglo-Protestant ancestry, which was challenged earlier by Catholics, Jews, other European ethnics, and the African American descendants of slaves. A society so constituted, and in possession of a strongly universalist mythology of the nation, confronts a striking circumstance: its national community—the "we" that corresponds to American citizenship—mediates more directly than most other national communities do between the species and those varieties of humankind defined in terms of ethnoracial affiliations.[56] A Chinese ethnic can, of course, be a citizen of France or of Great Britain, or even of Israel or Japan, but I hope it is fair to observe that, in these cases, he or she will encounter a national community with a manifestly more ethnocentric social history and public culture than that of the contemporary United States. Moreover, when this Chinese ethnic, or a white southerner, or any other American rooted in any one, particular enclave within the United States manages to identify with the American people as a whole, that American takes a tiny but ideologically significant step toward fraternal solidarity with the species.

To recognize this feature of American nationality is neither to deny that some other nations are comparably multiethnic nor to minimize the reality of ethnoracial prejudice, discrimination, and violence within American society. Nor is it to reinscribe Melville's romantic manifesto: "we Americans bear the ark of the liberties of the world."[57] One can easily enumerate the frauds and failures of the American effort to "share" American liberties with the world and to guarantee these liberties to those of our own citizens lacking the privilege of Anglo-Protestant ancestry. Indeed, so conscious are "we"—an alterity-preoccupied, deeply anti-imperialist generation of American intellectuals—of American arrogance, and so justifiably appalled at the uncritical enthusiasm for American military power displayed during the Persian Gulf War of 1991, that we tend to avoid earnest discussions of American nationality out of fear that the topic itself can yield only chauvinism.[58] But it is not chauvinism to insist that the ideological resources of the United States are too useful to democratic egalitarians to be conceded to the far right while the rest of us devote our public energies to more narrowly particularist or more broadly universalist projects.[59] A postethnic perspective invites critical engagement, with the United States as a distinctive locus of social identity mediating between the human species and its varieties—and as a vital arena for political struggles the outcome of which determine the domestic and global use of a unique concentration of power.

Such an engagement with the American nation need not preclude other engagements, including affiliations of varying intensity and duration defined by material or imagined consanguinity. A virtue of the term *postethnic* is to distinguish the perspective on American nationality sketched here from any reversion to a *preethnic* perspective on that nationality, according to which the general problem of the ethnos is dismissed rather than critically addressed and the specific issue of ethnoracial identity is suppressed by a monolithic, "100 percenter" notion of American citizenship. A postethnic perspective recognizes the psychological value and political function of groups of affiliation, but it resists a rigidification of exactly those ascribed distinctions between persons that various universalists and cosmopolitans have so long sought to diminish.[60] Angela Davis exemplifies this spirit of resistance when she asks that the "ropes" attached to the "anchors" locating individuals in primary communities be long enough to enable people "to move into other communities."[61] Welcome as is the cultivation of difference against the conformist imperative for sameness too often felt in American society, that very imperative for sameness can all too easily be reinscribed, in yet more restrictive terms, within the cultures of smaller, particular communities. Postethnicity projects a more diverse basis for diversity than a multiplicity of ethnocentrism promises to provide.

A postethnic perspective recognizes that the development of group identity takes place in the context of class position and that ethnoracial identity has often turned out, historically, to be a step toward, rather than away from, more complete participation in American life. The ethnic particularism of recent years amounts, in part, to another episode in the traditional politics of ethnoracial identity in America, by which persons outside the traditionally Anglo-Protestant majority have alternately asserted or diminished one special identity or another, depending on the apparent potential of that special identity to advance or retard one's fortunes in the society at large. A truly postethnic America would be one in which the ethnoracial component in identity would loom less large than it now does and in which affiliation by shared descent would be more voluntary than prescribed. Although many middle-class white Americans can now be said to be postethnic in this sense, the United

States as a whole is a long way from achieving this ideal.[62] This ideal for the American civic community is, indeed, just that, an ideal, embodying the hope that the United States can be more than a site for a variety of diasporas and of projects in colonization and conquest.[63]

Civic, moral, and epistemic communities differ so much that only the most general of observations can derive from any effort, like this one, to address all three within a single set of terms. But applicable equally to all three is Fernand Braudel's dictum that from "the question of boundaries" all other questions flow: "To draw a boundary around anything is to define, analyse, and reconstruct it."[64] If the truth in this aphorism is old, our own time is surely one in which we take special pride in the depth and range of our awareness of this truth. Our generation is perhaps unprecedented in the brilliance with which it creates good arguments for enclosures and exposes the naivete of efforts to avoid circle drawing. To recognize this virtue in ourselves may render us more critical in its exercise and may better enable us to facilitate the "Edwin Markham effect," according to which persons who are excluded help redraw the circles that enclose. The entry of women into the leadership of civic, moral, and epistemic communities is a salient example of this effect now in process, with an extent and significance still to be determined. But not all exclusions are bad, the conventional wisdom of our time will be quick to remind us, and we are all left with the responsibility for deciding where to try to draw circles, with whom, and around what.

CHAPTER TWO

A Historian's View of American Social Science

Dorothy Ross

THE EFFORT of historians to look across disciplinary lines at the social sciences is not unusual — historians often act as generalists. But due to the development of historicist and linguistic critiques of knowledge over the past twenty years, such cross-disciplinary study has become a more radical enterprise. In the 1960s and 1970s, a powerful critique of the academic disciplines that was mounted from the political left questioned the objectivity of disciplinary knowledge and the beneficence of specialized expertise. The social sciences were particularly vulnerable to this attack. Mainstream social science, taking natural science as its model, had made very strong claims to objectivity based on the use of the scientific method. The social sciences were also closely attached to the policy and helping professions, whose self-interested elitism was under attack.[1]

This political onslaught, as it turned out, was only the warmup to a more serious challenge. Theoretical developments from philosophy, science, linguistics, anthropology, and literature — from everywhere it seemed — were coming together to demonstrate the radically conventional character of all knowledge. In the words of Ludwig Wittgenstein, our truths are governed by the conventional rules of our language; the certainties we are able to attain rest only on the "customs," "forms of life," and "regular ways of behaving" in which our language is embedded. Although Wittgenstein did not say so, later theorists of the linguistic turn have recognized that to anchor knowledge in forms of life and regular ways of behaving is to anchor it in history. Indeed, they often call this position *historicism,* for to deny any fixed foundations to knowledge or value is to ground them in the historical world of meanings that human beings construct for their own purposes.[2]

Intellectual disciplines are preeminent examples of bodies of knowledge based in linguistic convention and social practice. The conventionalist critique invites us to examine the discursive strategies—from logic to metaphor—that govern each discipline and to locate their practice in the differentiated purposes of society and history. Specifically, each of us, rooted in one such tradition, is invited to look over our conventional walls at others from whom we have been deeply—but only historically—separated and to call into question the fundamental standards by which we work. My innocuous-sounding title, "A Historian's View of American Social Science," is meant to link cross-disciplinary effort with that large critical enterprise, one which is really just beginning.[3]

From my historical vantage point, then, what is most striking about American social science is its emulation of natural science and its liberal ideological boundaries. My readings in the social science literature, the left critique of objectivity and professionalism, and my awareness of the discursive limits of disciplinary traditions all lead me to regret the historical vacuousness and narrow political vision of so much of American social science. My standard of judgment lies in my understanding of historical practice and American politics and is informed by the more historically oriented and politically diverse traditions of social science that have developed in Western Europe. The alternative to a social science based in nature is one based in history; the alternative to liberalism is a broadened political spectrum that includes the organicist poles of conservatism and socialism. In the United States, mainstream social science transformed the historical world into a realm of nature and gave that realm the characteristics of an idealized liberal vision of America. In suggesting that the scientism and liberalism of American social science gave it a coloration different from that of Western Europe, I am not suggesting that it was altogether unique. The social sciences that developed in Europe also have ties to liberalism and positivism. American social science is one variant among others, a variant that accentuates the scientific aspiration and narrows the political range of a common, varied social science tradition.

How and why did this happen? The question points to the nexus between history and historicism, on the one hand, and nature and scientism, on the other, and to a long and rich discussion about the relations between these two modes of discourse. In no trivial sense, the social sciences were *about* history, and they constantly struggled with the problem of how to construct a science of the historical world. That struggle began with the eighteenth-century origins of the social sciences in the exemplary works of Adam Smith, Condorcet,

Herder, and others. The social sciences have generally been presented as extensions of Enlightenment science, and indeed they were. But they were also extensions of Enlightenment historicism.

Historicism is a mode of reflection about human affairs that developed over the course of several centuries; its implications are still being determined, as when contemporary philosophers link it to linguistic theory. Historicism means the understanding of history as a process of qualitative change, moved and ordered by forces that lie within itself. That view of history came into existence only when the human world was separated from the eternal realm of divine action. Only when loosed from God's eternal plan could people recognize the order that lay within human affairs and the novelty, contingency, and uncertainty that human action produced. Historicism began in the recognition by Renaissance writers that their own age was qualitatively different from the Middle Ages that preceded it and capable of reenacting for itself the ideals of the ancient world. By the eighteenth century, Enlightenment rationalism had reoriented historical time toward future progress. Modernity was now designated a novel course of historical development whose character was still unfolding and whose future was uncertain. The social sciences originated in this discovery of modernity. Its founders took—and their successors still take—as their problem the character and future of modern society.[4]

That eighteenth-century moment is interesting precisely for its joining of historicism and science. Adam Smith, for example, influenced by the work of Isaac Newton, propounded simple abstractions that would link qualitatively different kinds of phenomena and have the status of natural laws. Both as explanation and as a source of norms or values, Smith seemed to look to "the simple system of natural liberty." But historians have also found a prominent historicist strain in his thinking, which positivist traditions of social science overlook. Smith's central purpose in *The Wealth of Nations* was to demonstrate the idea of progress, and his economic analysis was framed by that purpose, not only in his historical account of the development of modern Western society in book 3, but throughout. Moreover, it is arguable that Smith found his norms, not in pure nature, but in imperfect history. Most often, he showed the market to be merely superior to mercantilism, not a perfect expression of nature: it was the experience of history that recommended the market to him. As was characteristic of the Enlightenment founders, Smith assumed a congruity between natural law and historical progress, despite his own recognition of the force of historical contingency.[5]

During the nineteenth century, both historicism and science developed in more self-consciously divergent directions. Under the influence of romanticism, historicists grounded reason and value fully within historical experience. The understanding of human affairs that historicists developed required that analysis be framed by unique historical configurations—stages, epochs, peoples, states, institutions, persons—that had the character of "historical individuals." Under the influence of positivism, on the other hand, theorists of science declared that the abstractive method and lawful structure of the natural sciences was the model for all fields of knowledge that aspired to certainty.

Despite this divergence, the social sciences remained in various ways and to various degrees involved with both their inherited traditions. Social science was not a unitary construct but a collection of partially distinct intellectual traditions, and some of these traditions stood in closer relation to historicism than others. The tradition of political economy begun by Smith, while not entirely abandoning history, turned in the direction of abstract science. Political science, however, drawing on national traditions of reflection about law, politics, and history, was closely linked to the historical study of politics. Nationality, too, was a powerful differentiator in the social sciences. In Germany, the Enlightenment and the romantic movement were closely intertwined, so that historicism developed there earlier and more fully than in the other European countries; German social sciences were historical studies.

There was also a more subtle and pervasive link between historicism and science. The Enlightenment's confidence in the congruity between natural law and historical progress was not so easy to maintain in the nineteenth century, when the understanding of historicism deepened and social conflict polarized opinion. The recognition of novelty in human affairs, for all its hope of earthly emancipation, aroused profound anxiety in cultural groups only lately withdrawn from the assurance of divine purpose in human affairs. The social sciences provided influential models of how that anxiety might be eased through the stabilizing effects of scientific law. John Stuart Mill, Auguste Comte, Karl Marx, and Herbert Spencer showed, in different ways, that modern Western history was propelled along a progressive course of industrial development by laws at work within history that were discoverable by science.[6]

So closely intertwined were naturalistic conceptions of law and historicist conceptions of change, that Karl Popper idiosyncratically—and to the confusion of scholars ever since—called that potent historical-naturalistic hybrid "historicism." Popper concluded perceptively that "it really looks as if historicists were trying to compensate themselves for the loss of an unchanging world by clinging to the belief that change can be foreseen because it is ruled by an unchanging law."[7] The compensatory function that Popper located in his social science theorists—the desire to evade or control the uncertainty of historical change by the imposition of scientific law — seems to have been at work in much of social science during the nineteenth century, and particularly in America, compensation was an especially powerful motive into the twentieth century.

In the United States, from the outset in the early nineteenth century, the variants of social science that began to develop employed the metaphors of nature or muted the attachment of their subjects to history. Still, American social scientists had some ties to historicism, and after the Civil War historicism became a more powerful theme in the culture and the social sciences. It was not until the 1920s that the mainstream of all the social science disciplines were captured by scientism: the self-conscious determination to follow the model of the natural sciences, a determination based on some version of the positivist belief that science offered a privileged access to reality.

One revealing example of this American path to scientism is provided by sociologist Robert Park. More than most social scientists, Park was concerned with the problem of how to understand the distinctions between history and natural science. This was partly because he took his doctoral degree in 1904 in philosophy, studying with Hugo Münsterberg at Harvard and then Wilhelm Windelband at Heidelberg, both of whom alerted him to the problem of historical knowledge. The natural sciences were then challenging the authority of historicism in Germany, provoking an effort to differentiate and legitimate the *Geisteswissenschaften* — the sciences of mind or spirit — and the *Naturwissenschaften* — the sciences of nature. Windelband argued that these two branches of knowledge differed only in their cognitive goals and the logical stance taken toward their subject matter. What characterized a natural or nomothetic science was its search for the general in the form of laws of nature. The historical, or ideographic, sciences, in contrast, sought the particular or concrete in the form of what is historically determined. Park adopted this distinction, and when he started teaching at the University of Chicago, he used it as an enabling charter for sociological science. Sociologists studied the same subject matter as historians did but took events "out of their historical setting . . . out of their time and space relations . . . to emphasize the typical and representative." Any subject that could state facts in that way and verify them was "so far as method is concerned, a natural science."[8]

But Park was not really content with this logical distinction between scientific and historical inquiry. He understood the general processes at work in history as natural phenomena and gave them the necessary and normative force of traditional concepts of nature. When Park urged sociologists to study the city, for example, he called the city an "institution" and then declared that institutions were "a product of the artless processes of nature and growth." "The city, particularly the modern American city," Park said, was created by "the inevitable processes of human nature," by which he meant the economic and social choices of competing individuals, "our system of individual ownership. . . . In this way the city acquires an organization which is neither designed or controlled." Later, he made that insight the basis of his ecological theory of society. "The economic organization of society, so far as it is an effect of free competition, is an ecological organization." Thus Park wrote the natural history of social phenomena, posited natural types, and mapped the movement of assimilating immigrants out from the center of the city as an ecological process. In this way, he turned the historical, institutional city into an inevitable product of nature and reified the economic and social processes of capitalist America.[9]

It happens that about the same time that Park was taking his degree from Windelband in 1904, Max Weber was using Windelband to clarify his own conception of sociology's relation to history and natural science. Weber accepted Windelband's distinction between nomothetic and ideographic studies, but unlike Park, he classed sociology with history rather than with the natural sciences. According to Weber, the cognitive aim of the sociocultural sciences was, like history, to understand concrete phenomena, the "meaningful and essential aspects of concrete patterns" and their "concrete causes and effects." Weber agreed that a natural science of society was logically possible, but he believed that it would be pointless. A natural science would give us only abstract representations of the features common to all events; we could not deduce from such formulas the "world historical" events and cultural phenomena that we find significant. Still, all conceptual knowledge, he argued, rests on generalization, and within the domain of ideographic, historical knowledge, generalization could be used to capture the concrete patterns of history. Hence, Weber developed a notion of ideal types that were understood as historical, rather than natural, constructions.[10]

If we ask why Park and Weber, starting from the same logical theory, diverged so sharply in their conceptions of sociological science, we can find many reasons, ranging from their logical abilities to their educational training. But perhaps the most fundamental reason was the different nexus between history and science that prevailed in the different traditions of discourse they inherited. Weber was raised in a culture in which, for more than a century, the unique configurations of his national history had been set off against and exalted above the recurring laws of nature. What he saw as significant in the sociocultural world were thus the concrete events of history, not natural laws.

Park was raised in a culture in which for more than a century the unique con-figurations of his national history had been identified with the recurring laws of nature. He thus subordinated the particularities of history to, and conflated them with, the regularities of nature.

This brings us to the heart of the problem. Why, in America, was history conflated with nature? The determining factor was the national ideology of American exceptionalism. American exceptionalism is generally portrayed as a kind of national myth, one that began in the exalted language of the Puritan "city upon a Hill" and that today often degenerates into chants of "America is number one." Indeed, the mythic idea of America was born in Europe, when inhabitants of what became the "old" world turned their imaginations upon the "new" one. This mythic America has been given many different con-crete forms: think of Martin Luther King's American Dream; or the immigrant dream of success; or the American mission to make the world safe for democ-racy. The background to all these versions of our national mythology, however, is a belief that America occupies an exceptional place in history.

The belief in American exceptionalism was formed at the time of the American Revolution and the decades following, intertwining Puritan ideas of America's holy mission, the republican hope that the American Republic, unlike all previous republics in history, would last forever, and liberal expec-tations of economic opportunity. The successful establishment of republican institutions and the liberal opportunity guaranteed by a continent of unculti-vated land appeared to set American history on a millennial course, guarded by divine providence. Embedded in this national self-understanding was a distinctive historical consciousness. American exceptionalism was prehistori-cist: it tied American history to God's eternal plan outside of history. Richard Hofstadter once remarked that America was the only country in the world that was born perfect and continued to progress. American progress would be a quantitative multiplication and elaboration of its founding institutions, not a process of qualitative change. As "nature's nation," and as God's, America could be seen as the fruition of the long, unchanging Anglo-Saxon lineage, or as the continuing domain of eighteenth-century natural law, where the course of American history followed the prescriptive norms of nature.

American exceptionalism also defined a particular kind of political econ-omy. Malthus's calculations might hold true for an old country like Britain, but in America, it was believed, vast stores of fertile land guaranteed that pro-duction of food would outstrip population growth. By the 1840s and 1850s, railroads, mechanized industries, and immigrant labor were clearly visible, but as long as western lands remained available for settlement and kept wages high, exceptionalism promised that economic independence would be widely available and wage labor need be only a temporary condition in the life cycle of the industrious workman. In America, the liberal market would distribute its benefits widely and hold the class conflict of European history at bay. The hand of time would be stayed, sparing America the fate of all previous republics: corruption, decay, and decline.[11] The belief that America, unlike Europe, was pursuing a utopian course of historical development thus stood at the center of an interwoven body of ideas. The exceptionalist idea of America quickly became the basis of nationalist ideology, entering into popular literary, religious, and political discourse. Indeed, from their earliest moments, social science writers took as their central task the maintenance of America's exemption from the fate of modern Europe.

That the social sciences focused on the problem of American exceptionalism is not really surprising. Historians have always known that Americans were intensely self-conscious about their national existence, so that we might well expect social and political thinkers to explore and defend the national self-conception, just as did American fiction writers and politicians. Moreover, American social scientists shared the national ideology not only as Americans but also as social scientists. American exceptionalism was one variant of the discussion about modernity within which the social sciences formed. The belief that Americans could embrace modernity, yet exempt themselves from its ills, emerged in part from a dialogue with Montesquieu, Adam Smith, and the contemporary discourse about the problematic future of modern society. Social scientists from the outset were involved in the set of issues around which both their intellectual traditions and national identity turned.

Exceptionalism, then, was the discursive frame within which the social sciences worked, the language that set their core problem and shaped the logic of their solutions to it. The role of exceptionalism in the social sciences was not merely rhetorical but deeply rhetorical. This is also to say that its role was deeply political and that the social sciences themselves were, in consequence, deeply political. This conclusion emerges as much from the work of Max Weber as Michel Foucault. Weber argued—in a manner congenial with contemporary linguistic theory—that concrete phenomena like that of history could be characterized in an infinite number of ways. It is the questions we ask of concrete social life and the interests and values embedded in those questions that select and shape the concepts we form. The university may give us some distance from the demands of practical politics, but it cannot divorce our reflections upon social life from our interests and values in it.[12] The language of American exceptionalism was constructed around a particular set of political attitudes and values that necessarily inclined American social science toward a liberal-republican construction of the social world and a deep suspicion of historical change.

This is a structural argument, but a discursive structure is itself a part of history. If exceptionalism fixed a certain kind of inclination, it was also subject to the mutations of history. The key components of exceptionalism—historical consciousness and political economy—are thus the keys to how the social sciences change over time. If we follow how attitudes toward history and political economy are transformed around the problem of American exceptionalism, we can follow the transformations in American social science.

At the time of its early nineteenth-century origins, American social scientists sought fixed laws of history and nature that would perpetuate established national institutions. Thus, Henry C. Carey rewrote the Malthusian and Ricardian laws of classical economics to guarantee American progress, and Francis Lieber linked his Kantian and historical principles of republican governance to what he called America's distinctive "calling."[13] But it is a long historical road from Carey and Lieber to Robert Park. The national ideology is itself a part of history and its vicissitudes.

There are two fulcrums in time around which this story turns, two periods of crisis that reoriented both the language of exceptionalism and the social sciences. The first was the Gilded Age, which I date roughly from 1877, the year of the violent, nationwide railroad strikes, to 1897, the end of the great depression of the 1890s, a period of marked economic and social upheaval. The second period of crisis spanned roughly the second decade of the twentieth century, encompassing the brief burst of Progressive reform hopes, World War I, and their disillusioning aftermath. Each of these historical crises moved the social sciences perceptibly closer to scientism and the metaphors of nature.

During the Gilded Age, for the first time, the exceptionalist faith was severely challenged. The weakening of religious belief loosed American history from divine protection. At the same time, rapid industrialization and the rise of class conflict forced Americans to face the possibility that America would change, that our own history would follow the same course that Europe had, and that permanent classes might develop. The threat to exceptional historical identity made the issue of historical change central, galvanizing two generations of social scientists to meet its challenge.

The first generation established the social sciences—first political economy and historico-politics—as university disciplines, hoping to reconfirm the traditional principles of American governance and economy on the basis of modern scientific knowledge. Unlike the antebellum economists, for example,

Francis Walker, observing a degraded class of factory laborers, admitted that Americans could no longer expect divine providence to exempt them from the laws of classical economics. But he set about to revise the iron law of wages and to paint factory conditions in America more favorably, thereby salvaging the exceptionalist hope.[14] But such efforts could not cope with the deepening crisis. As class conflict, labor organization, and populist protest mounted, socialism even reared its head within the ranks of the younger generation of social scientists. Particularly in economics, students of the German Socialists of the Chair urged that economics use the resources of history to control and transform the capitalist economy and its classical, laissez-faire theory. Some of them even predicted the historical transformation of the American Republic into a socialist democracy. Socialism raised the prospect of historical change in America in its most disturbing form.[15]

The Gilded Age crisis and its socialist moment produced a reorientation of both American exceptionalism and the social sciences. The socialist threat was put down and the liberal-republican heritage was pruned of its republican belief that political liberty and economic opportunity had to be based on the independent ownership of productive resources. In economics, the first of the social sciences to respond to the crisis, it induced John Bates Clark to become a leading theorist of marginalist economics and to establish the neoclassical paradigm at the center of the profession in the United States. Clark had initially been attracted to historicism and socialism in the early 1880s, hoping that providence was transforming American society into a cooperative commonwealth. But the violence of the Haymarket riot in 1886 and the conservative reaction it precipitated convinced him to change course. He developed a utilitarian theory of value—a core insight of marginal economics—in an effort to show that the market was a beneficent social organism.

Marginalism also allowed Clark to reconstitute the timeless domain of American exceptionalism. He achieved his marginalist vision, he said, by putting "actual changes out of sight, intentionally and heroically." The classical economists had begun early in the nineteenth century to abstract their doctrine from the full complexity of history. But they had still retained as their chief categories the physical factors of production—land, labor, and capital— and the historical classes that owned them. Ricardo's law of diminishing returns was understood to reflect a historical process of cultivation across the earth. In the marginalists' neoclassical world, these distinctive historical features dissolved. All values were determined by individuals calculating marginal utilities in the interconnected markets of capitalist society. The vision of competing individuals, freely exercising their wills and maximizing their goals through market choice, tending continually through surface fluctuations to reach an equilibrium of "natural" values, was an ideal liberal world. It was also the American exceptionalist world writ large.[16]

The logic of American exceptionalism, since early in the nineteenth century, had allowed Americans to imagine their development as a uniform expansion through space, rather than a differentiated development in historical time. On their continent of uncultivated land, Americans were in perpetual motion yet were held within unchanging historical parameters. It was Henry Adams, for a time a participant in gentry social science and a writer steeped in the literature of American exceptionalism, who gave that conception its most memorable metaphorical form.

> Travellers in Switzerland who stepped across the Rhine where it flowed from its glacier could follow its course among medieval towns and feudal ruins, until it became a highway for modern industry, and at last arrived at a permanent equilibrium in the ocean. American history followed the same course. With pre-historic glaciers and medieval feudalism the story had little to do; but from the moment it came within sight of the ocean it acquired interest almost painful. . . . Science alone could sound the depths of the ocean, measure its currents, foretell its storms, or fix its relations to the system of Nature. In a democratic ocean science could see something ultimate. Man could go no further. The atom might move, but the general equilibrium could not change.

Here was the potent vision of millennial America: America as the ocean — an unchanging realm of nature that had left behind the structured, changing European past; America as the final repository of law and the end of history; America as dynamic atomistic movement contained within a larger stasis. And here too was the attraction of science to American exceptionalists in a naturalistic age: only science could plumb an ocean that was part, not of history, but of the system of nature.[17]

Significantly, John Bates Clark used the metaphor of the ocean extensively in his writings to describe the marginalist economy, with its "ideal level surface projecting through the waves." Like America, the marginalist economy was an ocean of perpetual atomistic movement held in equilibrium by natural law. Against the socialist and historicist threat, Clark restated the American exceptionalist desire to escape qualitative historical change in the form of marginalist theory.[18] I am not contending, of course, that marginal economics was a uniquely American phenomenon. But it is noteworthy that Germany, with its historicist culture and organicist politics, resisted marginalism. It is also noteworthy that marginalism meant to Clark, and to many Americans who followed him, that America could continue to be — and these are Clark's words — a "progressive paradise." And it is noteworthy, finally, that Clark was only the first of a number of American social scientists to echo Adams's oceanic metaphor.[19] I doubt that they copied Adams. Rather, the metaphor achieved currency by its aptness to the logic and values of American exceptionalism, a logic and valence that the social scientists were reproducing in their own works.

Clark's neoclassical economics, then, effected an escape from history. But the historicism that had emerged in the Gilded Age and taken such dangerous form in socialism did not simply disappear but managed to put down American roots. Many American social scientists could not, like Clark, put "actual changes out of sight." Americans' utopian self-conception had for so long veiled the transformative effects of the machine in the garden that Gilded Age industrialization precipitated in many a sense of radical discontinuity between past and present. The constant refrain of the Progressive Era was that new industrial conditions had rendered obsolete the eighteenth-century forms and nineteenth-century sensibilities within which Americans still carried on their social and political life. The sense of discontinuity led to a new appreciation of historicism.

In the light of historicism, then, many social scientists began to revise their conception of America's place in history and their own disciplinary traditions. Nearly all of them — Thorstein Veblen was a major exception — adopted a position of liberal historicism and joined American history to the stream of Western liberal history. American ideals, they now said, depended on the same forces that were creating liberal modernity in Europe: the development of capitalism, social differentiation, democratic politics, and science. The shift marked a fundamental change in the national ideology toward a more cosmopolitan view of American history, but in the end, exceptionalism was revised rather than abandoned. America was now placed at the modern forefront (or the quintessential center) of liberal change, and universal progress was cast in specifically American shapes, so that America retained its exemplary or vanguard role in world history. The utopian hope for a harmonious society survived, its fulfillment now cast into the future rather than into the past. And having entrusted this ideal to the uncertainties of history, Americans continued to fear decline and to ward off the specter of class conflict.[20]

Under the influence of this revised exceptionalist hope, Progressive Era social scientists often incorporated liberal historicism into their models of social science. Note, for example, E. R. A. Seligman's *Economic Interpretation of History*, Edward A. Ross's *Social Control*, and Frank Goodnow's *Politics and*

Administration.[21] All of these works, however, display a deep ambivalence about history; it was often not clear whether their authors were writing about history or nature. And Ross's *Social Control* articulated the compensatory theme that would become increasingly important over the following decades. Precisely because America was now a part of history and open to change, history must be subject to scientific control. In the Gilded Age, it had been the conservatives — Clark in economics, Franklin Giddings in sociology, Abbott Lawrence Lowell in political science — who had championed science as a realm of unchanging natural law. Beginning in the Progressive Era, it was the reformist liberals, influenced by historicism, who now turned to science, hoping that knowledge of scientific laws would allow them to tame historical change.[22]

While Progressive social scientists were struggling with these contradictory impulses to evade, enter, or control the uncertainty of history, Arthur F. Bentley contrived to move another segment of American social science onto the American ocean in his book, *The Process of Government* in 1908.[23] Bentley envisioned society as an aggregate of individuals joined in innumerable associations and groupings, so that each individual participated in crosscutting affiliations. With this, he argued for the classlessness of American society. Despite what the socialists say, he insisted, there are no real classes "in great modern nations." No one, he said, has ever proved "that this hard grouping exists." What was true of modern society generally was especially true of America. In the United States, where group process was well developed, "groups are freely combining, dissolving, and recombining in accordance with their interest lines." This fluid liberal society thus produced a pluralist politics. Moreover, the group interests of American politics were constantly "harmonizing themselves," working themselves through to an "equilibration of interests." Here was, in form if not in name, the American ocean, the freely moving atoms of American liberal society, now joined by the freely acting interest groups of the American polity, held in equilibrium by scientific law.

Bentley's *Process of Government* began the second period of crisis that is the second fulcrum of my story. Again, as in the Gilded Age, changing attitudes toward history and politics led to a reorientation of the social sciences, this time pushing them into overt commitment to scientism. Bentley was representative of a new generation of American social scientists who felt alienated from Progressive politics and were determined to master the rapidly changing historical world that had so suddenly come into existence in the United States. Having left his academic post for urban journalism, he doubted his predecessors' sanguine assumptions of moral progress. After the trauma of World War I and postwar reaction, there were substantial additions to the ranks of those who felt remote from the American past, alienated by its moral politics, and skeptical of easy progress.

Alongside this moral and political alienation, and contributing to it, was a new sense of historical time. We have already seen how historicism and industrialization had opened up a growing sense of discontinuity between the American past and present. By the second decade of the century, the rapidity with which historical transformation was leaving the past behind produced a deeper sense of historical discontinuity. As the young economist Walton H. Hamilton said, "Change was everywhere." America was in "perpetual transition." World War I and the cultural changes that became visible in the 1920s only strengthened the sense of accelerating change. Sociologist William I. Thomas was one of many who articulated it. "We live in an entirely new world," he exclaimed, "unique, without parallel in history." [24]

This sense of living in a present that is radically new, cut off from the past, and in "perpetual transition" has become widespread in twentieth-century Western culture. It was articulated in Europe by modernist intellectuals in the late nineteenth century and was given classic expression in the years just before and after World War I. In this modernist view, history remains a realm of novelty, as historicism teaches, but the thread of continuity between past, present, and future is broken. Modernist culture dissolves history into a sense of perpetual transition, in which the present moment defines experience. There was a striking resonance between the modernist sense of time and the American exceptionalist orientation to historical time. A dynamic American history (which constantly recreated the contours of liberal society) and the perpetual transition of twentieth-century modernity both defined an American space exempt from the differentiated development of historical time. In short, historicism came late to America and was soon blunted by modernism, drawing social thinkers toward a vision of American history as perpetual liberal change.[25]

Modernist historical consciousness was reflected not only in Bentley's pluralistic behaviorism but also in other work of the 1920s. William F. Ogburn, for example, depicted social change as a repetitive series of "cultural lags" — culture always lagged behind material change and had to close the gap by adjustment to it. Ogburn's cultural lags; Robert Park's cycles of competition, accommodation, assimilation; William I. Thomas's process of organization-disorganization-reorganization; Wesley Clair Mitchell's self-righting business cycles — all these cyclical processes broke up the continuity of the progressive, historicist time of the nineteenth century into the perpetual transition of modernist time — and continually restored the ideal lineaments of American society.[26] The new experience of history also led American social scientists to focus their efforts even more strongly on the scientific method. During the war years and after, feeling more deeply alienated from the past and adrift in the current of perpetual transition, they began to develop a hard-edged, technocratic emphasis on prediction and control. As Thomas said in 1918, "This demand for rational control results from the increasing rapidity of social evolution." Once we have it, Thomas imagined, "then we can establish any attitudes and values whatever."[27]

Right after the war, sociologists and political scientists launched a concerted effort to refashion their disciplines around the rigorous use of the scientific method, by which they meant the quantitative method. The sociologists and political scientists reorganized the meetings and activities of their professional associations to encourage the use of quantitative methods. They jumped to support the Social Science Research Council, founded in 1923, which helped funnel an unprecedented amount of foundation money into the concern for scientific method and its promise of control. Neoclassical economists already believed themselves in possession of a scientific paradigm, but a new school of the institutionalists allied themselves with the sociologists' strategy.

The adoption of statistical methods in the 1920s opened new opportunities for the metaphor of the American ocean. The desire for control led American social scientists toward behavioristic premises and replicable, exact measurements. In that context, statistics became the hallmark of the new science. Beginning in the 1920s, many social scientists made statistical method the primary goal of social scientific practice, dictating both what was to be studied and how. Christopher G. A. Bryant aptly terms this American strategy "instrumental positivism."[28] It configures every object of study as an aggregate of individuals and every field of inquiry as the statistical study of aggregate behavior. In form, if not in name, it reconfigures the liberal, exceptionalist ocean, in which no structure impedes the flux of atomistic movement.

The patterns established during the early decades of American social science have continued through the twentieth century. The scientism that formed in the 1920s reached a higher point of development in the 1950s and revived again in the 1970s and 1980s. In each case, scientism, exceptionalist metaphors, and political retrenchment have been linked. And if American social science has taken conspicuous turns toward scientism in the 1920s, 1950s, and 1980s, it has taken equally conspicuous turns toward a revival of historicism and radicalism in the intervening decades of the 1930s and 1960s.[29]

The pattern follows the logic and confirms the existence of the exceptionalist frame of American social science. Most modern strains of socialism have been attached to historicist theory, so that a revival of radical politics has a conceptual link to historicism and triggers a call for historicism in social science, as it first did in the Gilded Age. In America, radical politics and heterodox social science methods often stand together, over against a mainstream social science that has incorporated the ahistorical, liberal, exceptionalist vision of America in its scientific practice and theories. The rise of heterodoxy, however, brings a reaction toward scientism not only from the right but from a liberal center that sees science as an escape from ideological warfare and from those on the left who believe that science can be used for critical purposes. Hence, the consensual pressure of exceptionalist politics, which dampens radicalism, at the same time strengthens the desire across the political spectrum for a "real" science, a "harder" science, that can place both the American ideal and social science inquiry beyond the reach of our wayward, controversial history.

Are the social sciences doomed to continue on this oscillating path? The answer would seem to depend on the continued strength of American exceptionalism in our culture and the continued cognitive authority of science in our universities. Both have been questioned in recent decades, but both are still powerful. During the 1970s, American exceptionalism appeared to be dying as a serious understanding of history, but the collapse of the communist empire has produced a strident revival of faith in America's unique world-historical role. Still, events may be conspiring to show us that the American past has in fact been more differentiated, and less ideal, than exceptionalist ideology allows and that the future of the world is likely to be, as well.

As to the authority of science, it now has a genuine rival in the new cognitive authority of historicist and hermeneutic approaches to knowledge. Some social scientists are already crossing disciplinary boundaries in the direction of the humanities, for example in historical sociology and politics, new kinds of institutionalism, and political theory. Still, the attraction of natural science remains very strong, and portions of the social sciences—like rational and social choice theory, sociobiology and cognitive psychology—have moved even closer to the scientific model over the last decade. Standing athwart the divide, the social sciences may well find their components moving in opposite directions.

There is a basis for bridging the divide in Max Weber's historical model of the social sciences. Unlike the exclusive privilege claimed by science, Weber's historical social science allows diversity. In the spirit of linguistic theory, Weber recognized that the historical field allows for different kinds of social studies, which ask different questions and therefore employ different methods and reach different levels of generality. Weber believed that neoclassical economics could continue under the aegis of historicism as one kind of ideal-typical analysis of the historical market. Market analysis rooted in history, however, would be aware of its social-historical limitations, reach out for historical connection rather than mathematical universality, and appreciate the possibilities of other kinds of economic study. This outcome, of course, would be satisfactory to a historian, but the social science disciplines, particularly economics, are still far from accepting it.

The social sciences have a great deal to gain from abandoning scientism for a looser historical definition of their field, for the conceptual and practical consequences of scientism have been misleading and counterproductive. A historical framework would force social scientists to recognize that their descriptions of the world are social, and inevitably normative, constructions that have great power to shape the discourse of modern societies. By and large, the language of American social science has been a reductionist language that encourages elite and technological manipulation rather than democratic self-knowledge and participation. It has also been a politically narrow language that tends to reify or ignore the structures of power in American society and, hence, to give us little critical purchase on our situation.

The current language of economics, for example, with its valorization of self-interest and its distrust of political activity, is corrosive of the fragile social ethic that survives in American political culture. A study by economist Robert Frank showed that twice as many economists as academics in other disciplines followed the logic of economic theory and gave no money to charity, although the proportion of free riders among economists was not large—about 10 percent. More graphic was his examination of college students who studied economics. Using the ubiquitous prisoner's dilemma game, he found that economics majors were significantly less likely to cooperate than nonmajors and that students were less likely to cooperate after taking an economics course than before.[30] Such evidence is admittedly meager, but it does support an intuitive sense that the reductionist discourse of contemporary economics is destructive of social values. By no means all of American social science merits that judgment. Some social science descriptions of the world have furthered egalitarian values and social change, such as the egalitarian racial theory of Franz Boas or Thorstein Veblen's analysis of the pecuniary culture of capitalism. But such redescriptions of the social world are not scientific in any special sense, and they are only mystified by being likened to natural science.

A historical model of social science would also revise the scientist's practical goal of prediction and control. For years now there has been a steady stream of criticism by social scientists and users of social science themselves, complaining about the gap between the abstract, sciencelike social science coming out of the universities and the complex, concrete needs of public policy.[31] Over the last decades, social science policy research has probably been getting more abstract, as microeconomic models have spread through schools of business and public policy. Following the incentives to scientize that are built into the academic disciplines, the helping and policy professions have tried to turn themselves into sciences and hence to approach practical problems through the model of prediction and control, thus reproducing the gap between scientific theory and concrete affairs that they were originally designed to close.

On the historical model, the people who study concrete affairs do not make scientific predictions; they make judgments. There are surely statistical techniques, mathematical techniques, and controlled experiments that might help one in making judgments. But judgments, by their very nature, require detailed knowledge of the problem in question, the kind of knowledge that comes from immersion in contemporary social practice and reflection on its historical and moral bearings. Social sciences that educated their students to make judgments would be giving them a very different education from the one they currently get, which is designed to turn them into as rigorous scientific methodologists as possible.

Despite, then, the difficulty of bridging the divided aims of the social sciences, I hope that heterodox social scientists will continue to break the lock that scientism has had on the social science disciplines in the United States and, finally, come to grips with history.

PART II

Medical Models

THE CHAPTERS in this section deal with the authority
of medicine as science and as a source of social, state,
and corporate power. JoAnne Brown considers the historic,
metaphoric interplay between contagion and criminality in
hygienic advertising, medical writing, and public health literature, suggesting the cultural permeability of each. Allan M.
Brandt focuses on the intersection of culture and science
while demonstrating the importance of how we think about
causation and responsibility in medicine. Regina Morantz-
Sanchez uses the career of Dr. Elizabeth Blackwell both to
explore a portion of the feminist critique of science and to
show how the marginalization of Dr. Blackwell on grounds
of gender deprived medicine of a perspective the power of
which we are rediscovering. Although differing in focus, each
of the essays raises questions about the autonomy, coherence,
and universality of "science." As Brown demonstrates, for
example, semantic relationships persist to constrain policy
debates long after their practical contexts are largely forgotten. Science was a product of choices, of exclusions, of
interpretation, as well as of research.

The stories told in these three chapters are filled with
ironies. In common with other authors, Brown assumes that
scientific authority is historically, socially, and culturally

contingent and that it does not rest on any absolute or universal validity. She explores the political implications of this assumption by identifying historical, analogic connections between hygienic propaganda and notions of lawfulness that cut across professions and their genres. Translations between hygienic and commercial domains that depended on martial and criminological analogies consolidated the informal powers of the state. The deep, historical purchase of these equations is preserved, however, in seemingly trivial bits of cultural ephemera. These small links among scientific ideas, commercial production, and the arms of the state grant science less autonomy but greater influence than do institutional analysis. Where Brown focuses on the cultural bearings of science as purveyed in historical metaphor, Allan Brandt's story is one of how an increasingly elaborate, specialized, and impersonal medical system came to focus on the individual and on individual responsibility, at the expense (as Brown and Morantz-Sanchez also note) of public health and a more social and environmental approach to health care. In Morantz-Sanchez's essay the ironies are ones of continuity and rediscovery. Her subject, Dr. Blackwell, had roots in the pre–Civil War antislavery and women's rights movements. In that sense, as well as in her skepticism about the germ theory and her appropriation of maternal imagery to justify her place in medicine, Dr. Blackwell looked backward to the nineteenth century. Yet she imagined a holistic, social medical practice that anticipated late twentieth-century feminist and other critiques of American health care. She envisioned—albeit partially and imperfectly—a medical science that would have met some of the criticisms raised by Brown and Brandt. The virtual absence of her voice from debates over the course of American medicine is evidence of the high scientific and social costs of pushing minorities and dissenters to the margins. It may also comprise a powerful case for locating science within the kind of pluralistic, historically self-conscious realm of social discourse argued for in the opening and closing sections of this volume.

CHAPTER THREE

Crime, Commerce, and Contagionism

The Political Languages of Public Health and the Popularization of Germ Theory in the United States, 1870–1950

Joanne Brown

Crime as Disease; Disease as Crime

I T IS 1923. Wealthy young scientist Charles Swinton has developed a new Scientific Theory to explain the evils men do. He has found Germs in the blood that dictate personality and cause the afflicted individuals to commit crime. He has devoted his life and wealth to finding a serum that will eradicate these germs and prevent crime, jealousy, alcoholism, and greed forever more.[1]

The story is fictional, of course: it is the script for a feature film produced in 1923, entitled *The Germ*, and written by Charles Swinton Warnack. But the twin ideas that crime is disease and disease, crime, have permeated the history of public health and personal hygiene in the United States since the 1870s.[2]

This metaphor has become on some levels so thoroughly incorporated into our way of seeing that it is hardly metaphorical, yet it emerged (or more exactly, reemerged) from a particular set of historical circumstances in which the theories, practices, and institutions of one enterprise—public health—became intertwined with another—law enforcement. At least in the American national context, these domains culturally reinforced one another, such that the diffusion of germ theories in public health amplified analogous contagion theories in criminology, law enforcement, and the political language of militaristic nationalism, and vice versa. That is, the criminological and military analogies of public health were not merely modeled after law enforcement and war but also became, and remain, models for criminology and military strategy. This cultural matrix, consisting of many variations on a theme equating disease with enmity, outlived some of the institutional arrangements that historically linked military and police force to public health. In this essay, I outline the half of this analogic system that pertains directly to hygiene: the role of criminological metaphor in the popularization of specific medical ideas and practices associated with the new germ theories of disease in the United States after 1870.

In the last two decades of the nineteenth century, ideas about "germs" reached a broad public, not only through the offices of regular medicine and the new public health but also through the burgeoning medium of print advertisement and through noncommercial (but also nonmedical) publicity, such as political advertisement. As cultural historians of medicine, health, and disease have shown, medical concepts do not live and die within the confines of medical institutions.[3] Public health journals, school textbooks, home medical guides, oral folkloric tradition, popular magazines, advertising ephemera, and motion pictures, radio, and later television all contributed to a popular imagery of germs as criminals or foreign enemies. Pictures and narratives from these diverse sources comprise the dramatic elements in terms of which we in the late twentieth century may understand sickness and health. A now-familiar, late twentieth-century "story" about sickness and health was defined, metaphorically, in the last third of the nineteenth century, when adversarial criminological and military analogies overtook religious and agricultural truisms about the nature of contagion, infection, and the mutual import of "seed" and "soil," inspired by Saint Matthew's parable of the sower.[4] The cultural persistence of the adversarial model for health and disease is striking when counterpoised against the dramatic material changes in health care in this century—and against the gradual decline in direct police enforcement of public health measures.

By the early twentieth century, cultural translations of state coercive power, commercially popularized in the imagery of the police and the military, persisted even as the state's exercise of the police powers of public health law and regulation waned. The concurrent "privatization" and decriminalization of public health, as represented by the work of prominent hygienist Charles Chapin, was matched by commercial appropriations of germ theory that nonetheless extended and transformed the power of the state even as public health authorities renounced their once-exclusive reliance on its police powers.[5] I argue that the diffusion of the imagery and language of the state's coercive power—the uniformed soldier or police officer, the courtroom, the notions of "arrest" and "protection"—through commercial hygienic advertisement operated to translate the rare actual use of state coercive power into a much more powerful, diffuse, and impersonal cultural form of authority.

In this translation of sheer power into cultural authority, the threat or possibility of coercion was a much more efficient and effective form of regulation, and self-regulation, than the messy business of rounding up large numbers of hygienic transgressors would have been. Exemplars such as the medical inspection of immigrants, or the incarceration of "Typhoid" Mary Mallon, or the deployment of uniformed health officers on city streets made this penumbra of informal cultural regulation meaningful, just as informal cultural regulation made these showcases meaningful. This effective cultural regulation was neither centrally coordinated nor necessarily purposive, yet it can be seen in retrospect as nonetheless systematic, in that much of the spontaneous cultural invention, the semantic creativity that produced the images of policed hygiene, depended for its apprehension on simple metaphoric equations between body and state, between medicine and law, and between disease and transgression.[6]

The commercial propaganda of hygienic "law" in this period, furthermore, accompanied two apposite tendencies in law enforcement: the increased privatization and corruption of policing and the increased professionalization and militarization of policing. Moreover, hygiene's dual analogic plot, set in the context of both law enforcement and military struggle, has had cultural significance beyond the bounds of medicine and public health. The crime-disease and hygiene-war equations ultimately strengthened the cultural institutions of policing and the military (including the reservoir of public language available for political discussion); that is, these equations strengthened the coercive powers of the state even as direct state use of the police powers of public health temporarily receded.[7] These developments were analogous to struggles between organized medicine and commercial manufacturers of patent drugs and hygienic sundries.

Publicity in a World of Strangers

In the second half of the nineteenth century, "intimacy of place, experience and attitude became radically severed from mutual trust."[8] A widespread perception of strangeness, and yet of increased economic and social interdependence among strangers, was shared by both long-term residents and new immigrants. This perception resulted from the triple impact of urban migration, increasing population, and immigration.[9] The technologies and institutions of mass communication and transportation that fostered these attenuating changes also mediated them and attained new social, economic, and cultural importance as a result. The social technologies of advertising, publicity, and standardization that arose after 1880 both undermined and enshrined the sorts of local, social knowledge that no longer dominated market culture. Though economic in purpose, these modern technologies had significant cultural and political import in spheres not directly related to these economic purposes.[10]

As national and international markets for goods and services overtook or augmented local economies, and as people acknowledged their increased interdependence, a new publicity economy developed to mediate between increasingly distant producers and clientele. For medicine and the social sciences, this mediation took the forms associated with professionalization.[11] Professional organizations and discursive practices worked simultaneously to monopolize and popularize professional expertise, translating esoteric knowledge for the marketplace.[12] In the skilled and professional labor market, and in educational and juridical institutions, this translation took the form of standardized measures of human capacity, which began to function as individualized brand names, verifying the merits of the person in terms of a universal standard. In commercial spheres, this mediation took the form of advertising; among health reformers, publicity.[13] The people who worked in these interstitial, mediating enterprises — advertising, publicity, personnel certification — often tried to regain a sense of personal connection with a mass audience by maintaining the imagery or metaphors of a more intimate set of social relations that, in their professional comprehension, had been lost or displaced in the acceleration of modern time.

Publicity had a further significance during this period. As the most salient "service" arm of the advertising industry, public health publicity — propaganda about contagion and its prevention — served as a "progressive" ideological antidote to actual contagion, "ammunition" in the "war" on disease.[14] That is, advertisers saw their campaigns as extensions of public health or medical practice, as a form of "inoculation" against illness, both physical and moral. A brief history of soap advertisement illustrates these points.

Advertising until the age of rapid rail transportation had been a primarily expository enterprise, subordinated to the production of goods: "BUY STAR SOAP. IT IS CHEAP AND GOOD. Sold at Johnson's Store."[15] The industrial creation of soap as a mass-produced commodity was vital to the early development of advertising institutions and techniques. Soap was until 1880 a perishable product made from animal fats and could not easily be marketed at great distances from its production.[16] Homemakers made their own soap from household renderings or bought it from small local producers. As technological changes increased the shelf life and transportability of soap, increasing consumer choices, each producer's soap had to be represented as well as presented, explained, and differentiated from alternatives. With very little intrinsic difference between one soap and the next, increasingly autonomous advertisers produced increasingly elaborate claims to create distinctions among soaps: suggestive brand names, invented traditions, promised effects, surprising properties, and dramatic associations.[17] The commodity itself began to assume these manufactured differences: perfumes, coloring, packaging all reinforced the fictional representations generated by publicists. Among these representations were those publicity claims pertaining to increasingly homogenized standards of beauty, to racial difference, to class identification, and to health.

The personification of germs as criminals, enemies, and strangers, and of public health agencies and soap companies as the arms of hygienic law, created a dramatic, adversarial script for the publicity designed to sell this commodified public health. At the same time, these popular dramatic representations helped culturally to define and maintain specific social arenas for policing. Thus, the representation of distinction and difference in the advertisement of basic commodities was not merely a reflection or by-product of preexisting political animosity, or of somehow "natural" sociological or cultural antagonisms, but was also among the requirements of a competitive commodities market. Mass markets produced, through advertising, a differentiation of simple commodities such as soap, and this differentiation magnified social distinctions. Advertisers adopted and exaggerated preexisting racial, political, gender, and class differences, with a consequent diffusion of these representations throughout the popular culture. The market-driven production of difference in print advertisement, as it mediated esoteric bacteriological knowledge, spawned the popular anthropomorphic imagery of the enemy germ and its commercial hygienic nemesis, soap.

The American medical profession was somewhat slower than commercial producers in accepting the utility of new germ theories of disease when laboratory evidence began to substantiate these theories in the 1870s.[18] Among American physicians, prevailing etiological understanding linked illness to general environmental and hereditary factors: sewer gas, miasmas, filth, parental intemperance, and organic poisons.[19] Even where microbes were admitted to play a role in disease causation, American medical thinkers relied on older environmental theories to dictate practice, general cleanliness being the most essential.[20]

The earliest "germ" theories, so named after an agricultural analogy, postulated that specific organisms caused specific diseases, "as wheat produces wheat, and rye produces rye, each after its own kind."[21] This idea did not dominate medical thought until the end of the nineteenth century, however.[22] The notion of specific disease causation by living organisms was a medically radical idea, made sensible by the agricultural analogy. As commonplace as the logic of disease specificity is today, near the turn of the twentieth century the profession's acceptance of germ theories marked a major change in prevailing American etiological thinking, a change not altogether welcomed by significant members of the regular medical community. New York physician Austin Flint, author of the standard medical text, *Principles and Practice of Medicine*, remained skeptical of germ theories until 1881, when he admitted their validity only in the case of the specific diseases of anthrax and relapsing fever.[23] Among some of Flint's colleagues, even this view remained unacceptable, for it dictated "antisocial" medical policies and required that physicians believe in invisible creatures. In 1882 another prominent New York physician, tuberculosis specialist Dr. Alfred Loomis, protested, "People say there are bacteria in the air, but I cannot see them."[24]

By the mid-1880s, however, even as members of the medical profession debated these issues, some commercial advertisers chose to adopt the popular, "amateur" belief in microbes, ignoring the niceties of professional skepticism but, nonetheless, attaching a medical imprimatur to these avant-garde theories. An 1885 advertisement for Procter and Gamble Company's new Ivory soap featured a testimonial from one Dr. R. Ogden Doremus, M.D., L.L.D., attesting to the soap's purity: "I subjected various samples of the Ivory Soap to rigid microscopical examination. I find it to be free from any forms of animalcular or vegetable germ life, so cordially commend the Ivory Soap for its unsurpassed detergent properties and purity."[25] Apart from its metaphoric social implications, the company's stress on purity alluded to the standard use of fats from diseased animals to make cheap soap. Ivory was made from vegetable fat, which did not spoil and was not suspected of spreading disease. Another late-century entrepreneur, Texas-born patent medicine dealer William Radam, produced a universal "Microbe Killer" made from large amounts of water, tinged with red wine, hydrochloric acid, and sulfuric acid. By 1890, when American medical practitioners were approaching consensus in accepting the practical, clinical importance of "microbes," seventeen factories were producing bottle upon bottle of the Microbe Killer.[26]

On the whole, although commercial and medical corporations alike entertained debates about germs throughout the 1880s, commercial firms in general were quick to borrow laboratory medical science, with its "normal science" approach to bacteriology, while often playing down clinical physicians' pragmatic skepticism of germ theories. That is, the laboratory scientist became a more useful source of medical authority for commercial advertising purposes than did the general practitioner, although both represented medicine.[27] At the turn of the twentieth century, American manufacturers who used germ theory as a selling point for their products often strove for a scientific tone in their copy, as opposed to later, more melodramatic approaches. Such companies did not hesitate to instruct the public as to the significance of recent biomedical discoveries. The Sanitas company of New York produced a ninety-six-page brochure, "How to Disinfect," which included an explication of "the exact connection between Microbes and Disease." "Disease ensues," the manufacturer explained, "as a direct consequence of the multiplication of microbes," whereby the vessels of the body become "clogged" with germ life and its debris.[28]

The Lever Brothers Company claimed, similarly, that their Lifebuoy Royal disinfectant soap exterminated the "various germs and microbes of disease." [29] Charles Tyrell's "celebrated J.B.L. Cascade," a hydropathic system of internal bathing, purportedly rid the system of the "germs of putrefaction" that caused disease.[30] Such popular translations confused "ancient," "early-modern" and "contemporary" scientific ideas of disease causation, evoking mechanistic hydraulic models of human biology as well as filth and putrefaction theories, while acknowledging the importance of "germs" and "microbes." Any strict chronology of germ theory, in which one analogic paradigm fully supersedes another, vastly oversimplifies this picture.

During the first decade of the new century, such commercial advertisements were an important source of information about disease causation, one widely available to the general public and often, by design, indistinguishable from public health notices. In retail store pamphlets, on packages, and in both advertisements and articles in general interest magazines, copywriters struck an authoritative scientific tone. The Philadelphia manufacturers of Pratt's Germ-a-thol stated in their brochure, "It is a well-known fact many diseases are caused by germs entering the body, increasing rapidly and disarranging the entire system." In order to prevent disease, the company affirmed, germs must be destroyed before entering the human body. The manufacturers evidently remained confused about the exact sequence of causal relations: "It is this micro-organic life that causes decomposition, mould, mildew, filth, bad odors, and fungus growth in which germs, vermin and insects live and multiply by the millions." With this comprehensive, albeit tautological, warning, Pratt's Company recommended Germ-a-thol for "laundry, windows, carpets, furniture, bath, feet, mouth, scalp, douche, nasal catarrh, burns, lice, toothache, plant lice, poultry houses, stables, hogs, and dogs."[31] Germs and microbes were added to an already long list of environmental factors popularly understood to promote disease.

Manufacturers of disinfectant, soap, and patent medicine were soon joined by other entrepreneurs in mighty "battle" against the microbial agents of disease. A 1906 advertising campaign by the Gillette Company on behalf of its newly invented disposable razor blade used the imagery of laboratory science and bacteriology to discredit the old-fashioned, germ-infested (and owned predominantly by Italian Americans or African Americans) public barbershop, the "seed-bed" of "unmentionable diseases."[32] Microscopic views of skin, hair, clogged pores, dandruff, and bacteria became the stock-in-trade of advertising illustrators. These ersatz laboratory images reinforced the credibility of etiological theories based on otherwise invisible organisms. Such images, making vivid an unseen world of danger, played upon popular fears of the unknown and the invisible while seeming to rationalize these same fears under the aegis of science.

Once the germ theories of disease settled into the standard medical literature, by about 1890, the commercial use of germ ideas and imagery accelerated. After the turn of the century, an even wider range of commercial products found part of their appeal in protecting people against the depredations of germs. Notions of contagion and its prevention expanded in scope and were turned to new commercial, cultural, and political uses. Before about 1905, for example, producers marketed many household products in bulk, without individual wrapping of any kind. Manufacturers sold soap wholesale, in long bars, to grocers; the customer purchased it by the pound, the grocer cutting the desired amount from the common bar with a wire knife. Pharmacies sold toothbrushes unwrapped, displayed in large bins or baskets at the pharmacy, where the customer could test the quality of the toothbrush by running a thumb across the bristles. These customs began to change at the turn of the century, and popular germ theory was one urgent rationale advanced for these changes by packaging companies that stood to profit from them. The company producing the Pro-phy-lac-tic toothbrush distinguished it from all others by its individual, safety-colored red and yellow packaging and by advertisements that warned, "Do not buy from a fingered pile of dusty, germ-laden toothbrushes handled by nobody-knows-who."[33]

Although often unnamed and unspecified, the "nobody-knows-who" whose germs and hygienic ignorance posed such mortal threats to the cleanly were commonly—not surprisingly—portrayed as members of "alien" or stigmatized groups. In keeping with broader cultural conventions, advertisers, medical journalists, and public health reformers portrayed African Americans, Native Americans, and Jewish, Chinese, Irish, Italian, and other immigrants — and the poor in general — as inherently dirty and potentially dangerous; their suppression, conversion, or physiognomic transformation was among the many duties of the righteous and the cleanly.[34] These well-worn themes appeared in many early twentieth-century illustrations and advertisements. As racial insults, their pettiness may seem to render them unimportant, particularly in this era of extreme racial violence, but they are worth examining as well as condemning, for they have a more subtle significance that pertains to the militarization of hygiene and the aggrandizement of state authority.[35] Indeed, the very triviality of any single exemplar comprising this vast, uncoordinated, commercial culture makes it both difficult to study and difficult to resist, and in these very difficulties lie some of its political force.

In a children's book published by the makers of 20-Mule Team Borax, "black imps," of African physiognomy, threaten the health of a nursery room, until they are vanquished by the white uniformed soldiers of the 20-Mule Team Brigade.[36] Procter and Gamble, in a style of advertising argument common to other soap manufacturers, suggested that their soaps would help Uncle Sam in the Indian wars, by "civilizing" the "savage."[37] The German-American Lautz Brothers' soap company was among the many companies who portrayed the very skin color of African Americans as dirt, to be magically washed away with Lautz's lye soap.[38] In another popular advertising campaign, by Enoch Morgan's Sons on behalf of their all-purpose cleanser, a street sweeper used Sapolio soap to cheer himself up when his work was taken from him by zealously clean townspeople. Under the watchful eye of the Spotless Town policeman, he washed his "black" face "white" with Sapolio and was shown to be much happier for it.[39] In the dominant white culture, the person of color often occupied a liminal position: though fundamentally "unclean," he or she could be trained to acceptability by white European society upon thorough, miraculous conversion to whiteness, and all that entailed, through the magic of purchased hygiene. Variations on this message that emphasized color-coordinated levels of respectability were purveyed for African American markets, as historian Kathy Peiss has recently shown.[40]

Conversely, and not surprisingly, when advertisers personified dirt, germs, or microbes, these characters were almost invariably dark skinned and large featured — recognizably African, Italian, Slavic, or Jewish. Analogy tends to work in this reciprocal fashion: if poor people, people of color, and immigrants are stereotyped as dirty and diseased, then dirt and disease quite easily assume the faces of the poor, nonwhite, and foreign-born.[41] Tera Hunter shows how the caricature "Soap Sally" worked for white-owned commercial laundries in Atlanta to discredit the independent, African American washerwoman.[42] Though such crude stereotypes were most common in advertising before 1930, as late as 1951 a major toothpaste company portrayed dental bacteria-causing tooth decay as animated black "slaves."[43] What these figures have in common is their portrayal as enemies of the American people and of the American state, in their caricatured bacterial assaults on the body.[44] Although advertisers stigmatized different groups for different markets, at different times, the generic images of the germ as "enemy," and of hygiene as "defense," remained constant, with continuing implications not only for health, disease, and medicine but also for our political conceptions of social danger.

One of the most successful advertising campaigns of the new century was mounted, not in magazines or newspapers, but in streetcars, a significant locus of political contest before World War I.[45] Like the new packaging trend, this campaign on behalf of Sapolio all-purpose cleanser capitalized on the widespread distaste for contact with strangers, and nowhere were city dwellers more likely to rub elbows with strangers than on a streetcar. The streetcar, operating as a commercial and social "carrier," became a place for the transferral of social and medical danger.[46] It was a place where intimacy, or at least close physical proximity, was foisted upon strangers. The streetcar became a major locus for intensive public health propaganda and commercial hygienic advertisement; even the transfers in many cities bore "No Spitting" messages.[47]

In 1900 the makers of Sapolio all-purpose household soap, through the agency of Albert D. Lasker, created the imaginary Spotless Town, full of hygienic heroes, and Dingeytown, teeming with filthy would-be emigrés, whose escapades appeared serially on city streetcars—the original "soaps." Over the six years of the advertising campaign, real towns voted to become Spotless Towns, schoolchildren practiced for "spotless" school plays, and the campaign became a classic in advertising literature.[48] The hero in the Spotless Town drama was the policeman, who enforced cleanliness; he prevented immigrant Dingeys from coming across the border to commit their dirty crimes. A frightening tale was set to verse in fourteen parts:

Listen, my children, whilst I relate
The harrowing, hideous, horrible fate
When an inky flood over Shining State
The fair land threatened to inundate.

. . . .

On the jump came the guards at the first swift alarm,
To the arsenal speeding, to muster and arm
Brave men from the court-house, the cottage and farm
To shield the fair name of their city from harm.

. . . .

With their dark forces routed, in shame and in fear,
The filthy host fled to their hovels so drear;
Nor ventured again to this town to draw near,
Except when clean-living and light to revere.

. . . .

My children, you know, that you all want to grow,
To fight grime and dirt as we did long ago;
But of weapons you use, I want you to know,
There's none that will aid like SAPOLIO.[49]

In the service of selling soap, this allegory of invasion, civil unrest, and darkness met by an enlightened militia capitalized on nativist and white-supremacy sentiment rampant at the turn of the century.[50] Such popular imagery also fostered perceptions of "alien races" as malicious, perverse, lawbreakers and degraded criminals and reinforced cultural connections among immigration, filth, crime, and disease. Conversely, these advertisements, and many others based on the same plot, identified native-born white Americans with the force of law and with physical and moral cleanliness. Although the Sapolio campaign focused on visible dirt rather than on invisible germs, the connection between the two was strongly asserted in other contemporary pronouncements on hygiene.[51] Moreover, the Sapolio campaign greatly reinforced a triangle of cultural associations between contamination and immigration, between immigration and crime, and between crime and contamination. This reinforcement makes advertising significant politically and culturally: as "integration propaganda," such cultural forms take the world and make it more so.[52]

The criminological trope was a particularly useful one for representing sanitation and hygiene in familiar dramatic terms. The spectacle of the courtroom trial was familiar to theater and moviegoers, who may never have set foot in a court of law, and presumedly even more familiar to the "criminal element" targeted by many reformers who took these metaphors seriously. A publicist for the Wisconsin Anti-Tuberculosis Association argued succinctly, "to the intelligent mind open to facts of science, spitting is murder because it spreads abroad the germs of disease and death."[53] For reformers who saw legislation against sanitary vice as a solution to public health problems, the enforcement of health law, and the prosecution of infractions, was persistently imperfect.[54] The metaphorical depiction of the courtroom trial can be seen as a sort of graphic wish fulfillment.

Medical historian Naomi Rogers, in documenting the unmaking of the housefly, uncovers several such representations of the sanitary "culprit" on trial.[55] Other courtroom trials appear in the World War I–era film cartoon, *Health Twins at Work,* by the American Social Hygiene Association. In the film, tuberculosis, infantile diarrhea, and diphtheria are captured by the health twins, Science and Administration, and stand before the presiding judge, Public Opinion.[56] As scholars of immigration and nativism have established, the link between criminality and the non-Anglo-Saxon foreign-born was a social-scientific commonplace in this period. Yet it is important to recognize the intricate ways in which this connection was made. Medical experts did not merely respond to preexisting nativism but participated in its explication en route to catechizing the public about hygiene.

The notion that all immigrants were disease carriers, invading "hosts" to dangerous germs, and likely to flout hygienic law, was fostered not only by xenophobic advertising but by news accounts of medical inspections of immigrants arriving at Ellis Island. The medical inspection of immigrants had become part of the U.S. Public Health Service's mandate in 1893, under the leadership of Dr. Walter Wyman, as an extension of the Health Service's quarantine responsibilities. About the same time, officials across the country introduced similar examinations of schoolchildren.[57] Some immigrants did suffer from contagious and infectious disease and were deported as a result, although the percentage was small, never exceeding 3 percent of arrivals at all U.S. ports between 1898 and 1916. Interestingly, however, medical exclusions actually increased over time in proportion to other exclusions, including exclusions for criminality.[58] In a small but symbolically significant way, then, medical judgment came to replace other forms of expertise in regulating the entry of foreign-born people to the United States. Medical disqualification accounted for 69 percent of rejected immigration applications by 1916, according to one scholar's estimate.[59] Immigrants who were allowed entry into the United States, then, met a certain health standard to which few white native-born adult Americans were subjected.[60]

The photographs and writings of the U.S. Public Health Service contributed to popular associations between immigration and infection. The very existence of the inspections, much like the appearance of public health notices on streetcars, implied that these interventions were both necessary and effective. The inspections called attention to the bodies of immigrants and created a role for selective, intimate, intrusions by the state police powers of public health, consigning immigrants to the same subordinate status as schoolchildren.

The language of public health officials assigned to this duty could be harsh. In 1912 and 1913, Dr. Alfred C. Reed, an officer at Ellis Island, where more than two-thirds of all immigrants came ashore, wrote about "the medical side of immigration" in terms of political sanitation: "It can truthfully be said that the dregs and offscourings of foreign lands, the undesirables of whom their own nations are only too eager to purge themselves, come in hosts to our shores." [61] This discussion, in what was at the time a highbrow journal, *Popular Science Monthly,* recalled the language of the Sapolio streetcar ads, dehumanizing the immigrant with terms like "dregs," "offscourings," "purge" and "hosts," and projecting an American desire for purification onto other states. "The immigration current," Reed argued, should not be "allowed to stagnate in eastern cities." [62] In an era of widespread xenophobia, the two enterprises, one ostensibly commercial and one frankly political, shared an equation for expressing alarm: one dramatized filth with an immigration scenario, while the other dramatized immigration with a filth story. [63] In one case, soap was being sold; in the other, immigration was being restricted. In both, the solution to the problem was a uniformed officer — the Spotless Town police force or members of the U.S. Public Health Service. Reed's article, illustrated with a photograph of five health officers in military-style uniforms, announced that "the medical inspection of immigrants is the first, most comprehensive, and most effectual line of defense against the introduction of disease or taint from without." [64]

The conflation of the military and police categories, in the figure of uniformed health enforcers, was in keeping with this militarization of the police profession, yet it occurred even as public health policy turned away from the widespread use of police powers. How can this contradiction be explained? I argue that the nature of police power is such that its actual use is only the cutting edge of its full social force and that the threat of police power constructs an authority much broader than the scope of any specific laws, regulations, or enforcements. In this way, the language and imagery of enforced hygiene, backed by even an anachronistic uniformed health officer, is potentially as effective an influence on hygienic behavior as are mass arrests — and considerably less expensive. Moreover, as Judith Leavitt suggests in her recent work on "Typhoid Mary," such public health "show trials" (though no formal trial occurred) instruct citizens on the powers of the state, not only through their enactment as regulation but also through their enactment as social theater and their many translations into commercial culture. [65] The intercurrence of commercial culture into public health policy, then, could not be contained, any more than the corporate use of private police forces could be parceled off from the workings of state law enforcement and criminal justice.

The picture is further complicated by the cultural vestiges of a much older, European notion of "medical police," who from the seventeenth through the nineteenth centuries were supposed to be responsible for enforcing public health laws and quarantine, overseeing medical practice, and inspecting hospitals. According to mercantilist political theory, their health enforcement activities were crucial to maintaining a national population necessary to the strength of the early modern European state.[66]

This formal notion of a medical police force never fully materialized in either Europe or the United States, although medical inspections of immigrants by uniformed public health officers appear as vestiges of cameralism's ideal. And although legislators strengthened public health statutes in the early twentieth century, municipal police became progressively less involved with enforcing these laws as municipal health departments took over sanitary functions, gained some powers and duties previously conferred upon police, and like the police, adopted military-style uniforms.[67]

The uniformed health officer also appeared in a 1909 antituberculosis filmstrip, The Jolly Health Cops, in which a man was hauled off to jail for spitting on the street.[68] The uniformed police officer as a symbol of obeisance to hygienic law was not a mere figment of illustrators' imaginations and their equation of hygiene and law but also an obsolescent representation of health enforcement practice: it personified a history.[69] Moreover, the often interchangeable language and imagery of soldiers and police, law enforcement and war, reflects the historical convergence of these two domains — military and civilian enforcement — during the last half of the nineteenth century.[70] It was during this period that urban police and constabularies adopted military hierarchy and military uniforms in their self-conscious efforts to professionalize, partly in response to the incursions of private hirelings such as the Pinkertons.[71]

In health reformers' formal declarations and in legislation after 1910, an increasing emphasis on education (persuasion) replaced the use of police power (coercion).[72] Public health physician Charles V. Chapin explained: "Education has become one of the most important factors. . . . The physician and the nurse are the chief agents in the new movement. They have taken the place of the sanitary inspector and the policeman."[73] Nonetheless, the older image of the medical "officer" would remain in force in American hygienic advertising for the next half century, through the 1950s. This police image alternated and merged with martial images of human struggle against disease, continually reconfigured as the social and technological waging of war also changed.

I do not want to leave the impression, however, that metaphorical forms of representation are mechanically linked, in any fixed way, only to material and historical contingencies. Rather, I argue that both immediate experience and social memories embedded in verbal and visual languages are continually at work translating the news into extant vernacular, while past experiences are recast in light of the new. Private experience is continually translated into public discourse, and vice versa, as the past enters the present through borrowed language. Yet this conservative process also allows for change: since multiple analogic equations (past and present) coexist in any culture, and between cultures, new metaphorical combinations are available—the more so when geographic mobility, the practices of history, memory, and communication are freely exercised.

The "New" Public Health in a World of Strangers

In 1910, Dr. Charles Value Chapin published a treatise on the "sources and modes of infection," which lent considerable force to general fears about the spread of disease through personal contact with strangers.[74] Chapin described, in often disgusting detail, the unsanitary habits that foisted "the secretions of others" on the hapless modern urbanite. Much of the revulsion generated by Chapin's text depended upon, and further reinforced, negative perceptions of strangers. This small book, which became a best-seller, helped to shift popular ideas about infection away from traditional environmental concerns and toward the problems of personal hygiene. Chapin assured his readers, "It is usually comparatively simple to live so as not to allow the secretions of others to come into contact with one's own mucous surfaces."[75] But his graphic description of the everyday transference of "secretions" belied that assurance.[76]

Under the conditions of ordinary life and commerce in the early twentieth century, and according to prevailing bacteriological theory, the conveyance of germs from person to person, through unregulated food handling, inadequate sanitary provisions, and naive personal hygiene, seemed significantly more likely than in earlier years. In urban settings, germs seemed to spread more rapidly than scientific ideas. Urban food distribution was more complex and attenuated than rural; sanitary facilities were unreliable; medical knowledge was often restricted to professionals; and urban dwellers lived in ever more crowded conditions, constantly exposed to "nobody-knows-who."[77] The ordinary routines of daily life seemed fraught with danger, as contact with strangers became more likely and as popular acknowledgment of an invisible microbial world gave yet another scientific rationale to xenophobia, in addition to the eugenical one.[78]

In the folklore of popular germ theory, "communicable" disease microbes tended to favor new-fangled inventions as their habitats, the streetcar and the elevator being among these. Ironically, the public health department notices that papered these public places added to their peculiar stigma. The telephone, with its proximity to the mouth and ear and its conveyance of the invisible across a great distance, was a particular concern. New modes of communication fostered fears of new "communicable" diseases; this association was no mere historian's pun. The antituberculosis *Journal of the Outdoor Life* published this knowing account in 1906: "A Seattle matron, instructing a new housemaid in the duty of cleansing the telephone, was interrupted by the maid with the assurance that she fully realized the necessity for such a precaution. 'Why,' said she, 'once my sister lived in a family where they didn't know they ought to do that; and one day, when one of the ladies went to use the telephone, she found a *great big microbe right on it!*' "[79]

Older European traditions, including scriptural tradition, associating money and commerce with filth and disease returned in this period as part of the growing recognition of the contagiousness and ubiquity of tuberculosis. "Every time we men place a coin of filthy lucre between our lips we dare the white plague to do its worst," warned a Harrisburg physician in 1904.[80] This widespread and ancient concern resurfaced dramatically in the paper money scare brought about largely by one man's agitation, A. Cressy "Clean Money" Morrison.[81] Library books also worried some hygienists. From 1647 until 1968, mail in the Americas was regularly disinfected during epidemics of yellow fever, smallpox, bubonic plague, typhus, and cholera and during some outbreaks or concentrations of diphtheria, measles, mumps, tuberculosis, scarlet fever, influenza, and leprosy.[82] Books and letters (as "propaganda" of contagion) and money (as symbol and artifact of commerce) represented a particular hazard of the male public life, while clothing and laundry carried danger into the female private sphere: "Every time a lady brushes the pavement with her dress she creates the possibility of carrying the seeds of consumption into her own home."[83] In both instances, it was the possibility of indirect contamination from strangers' bodies that posed the threat to the white middle-class person and home.

Tuberculosis, the "white plague," was the single greatest disease threat in this period. Patients often shared the popular construction of themselves as hygienic criminals and their predicament as a form of imprisonment. By 1912, a diagnosis of tuberculosis legally subjected them to the police powers of the health departments.[84] Tuberculosis sufferers frequently referred to themselves as "cons," meaning not only the archaic consumptive but also the convict and conscript.[85] Often, the comparison was offered as comic relief: one sanitarium patient in Pennsylvania was overheard to say, "If I ever get my 'bugs' arrested, I'll try and have 'em tried before a jury and sentenced for life." [86] Another patient described "doing time in camp" for "seven months, thirty and two-sevenths weeks, 212 days, 5088 hours, 305,280 minutes, 18,316,000 seconds" without a "trip to civilization," only to find that the hectic pace of urban life made him grateful to return to the confines of the sanatorium. [87]

As metaphors tend to do, the disease-crime equation worked both ways, and this circularity ultimately defeats attempts to ascribe semantic cause and effect. [88] Historically and semantically, germs and crime are messily connected, and this messiness is belied by the neat symmetry necessary to historical narration. It is nonetheless instructive to note where the crime-as-disease analogy (the converse of disease-as-crime) appeared. In a 1906 address to prison physicians, tuberculosis expert S. A. Knopf explained, "Consider the prison a hospital for the morally diseased whose unkindled souls are clothed in bodies often imperfect by reasons of the sins of their fathers, often greatly debilitated by disease or privation, and then consider yourselves designated to be the directors in the treatment of these bodily and moral ills. . . . You may well feel that your calling of prison physician is a holy and sacred one." [89]

Just as legal historians and sociologists have found the law to be unequally applied, both the metaphorical indictments and the legal prosecutions of contagion fell differentially upon different classes of people. [90] Yet at the same time, the crime-disease equation magnified the ability of each (crime and disease) to ensnare in its institutional sanctions and cultural stigma the ignorant, the rebellious, the persecuted, the inept, the malicious, and the innocent *of all classes:* these equations, as forms of integration propaganda, also created a "class" of the sick and potentially sick, a category potentially more malleable than economic class, gender, or race. These categories, doubling one another and standing in for one another, expanded the police powers of the state not merely institutionally but culturally. The double-edged equation between disease and crime, crime and disease, broadened the notions of the "respectable" and the "disreputable" beyond class status or racial identification and allowed the state to impinge upon even the prerogatives of the privileged, even as that class promoted and directly benefited from the state's hygienic regulation of poor and disfranchised people. [91] Gender and racial discriminations, also inscribed in cultural metaphor, further destroyed any neat class orderliness, just as class always cuts across gender and racial lines, race across class and gender, and so on. [92]

The trite observations that "germs know no color line" and were "no respecter of persons" also carried the implication that germs, like twentieth-century persons, commodities, and information, were on the move. [93] This geographic mobility was not limited to the well-to-do. Visiting nurses and other public health officers readily saw the difficulties of maintaining contact with the physically mobile underclass. A series of economic depressions in the 1890s, the horrific resurgence of lynching in the South and the refugees from it who moved northward, and continued European immigration up until 1924 all contributed to internal migration and to contemporary observers' sense of the increased mobility of the poor.[94] In addition, travel to points south and west, prescribed between 1849 and the 1930s, remained among medical recommendations for the relief of tuberculosis, the nation's leading cause of death. Large towns and small cities in the Southwest were founded upon the migration and care of these "health-seekers," and a powerful backlash was evident in the Southwest after 1900.[95]

African Americans found themselves disproportionately charged with sanitary crimes, both legally and figuratively, during their years of enslavement, and Emancipation did not change this pattern; indeed, their northward migration seemed to exacerbate it even while being motivated by the violence that racked the South.[96] Centuries of white European abuse had constructed a justificatory notion of the Negro as subhuman beast; under the radical racism of the Reconstruction period, white paranoia imagined in black men an inherent viciousness that, beginning in 1890, ostensibly justified the unspeakable atrocities of lynching that seared the nation for more than twenty years.[97] The purported "Negro crime" of rape became yet another basis for public health reform rhetoric, as the author of a 1909 article in *McClure's* magazine, "The Vampire of the South," insisted: "Negro crimes of violence number dozens where his sanitary sins number tens of thousands. For one crime a mob will gather in an hour to lynch him; he may spread the hookworm and typhoid from end to end of a state without rebuke."[98] The infamous racist newspaperman John Temple Graves, of Atlanta, even suggested that a research foundation be established to locate the "germ of the rapist" that seemed to "infect" the black population of that city.[99]

Germ theory, as I argue elsewhere, evolved in tandem with the racist post-Reconstruction ideology of white supremacy and was consistent in many fundamental ways with racist fears of miscegenation and sexual pollution.[100] Germ theory inspired a versatile medical counterpart to Spencerian Social Darwinism, which I call Social Contagionism. Here, the older "seed" etymology is pertinent, making vivid the conceptual similarity among agricultural, sexual, and bacteriological usage.[101] Germ theory, in a sense, extended the implications of miscegenation, for it implied that casual contact, and not only sexual intercourse, could spread racial illness. African Americans were not alone, however, in being subject to this spurious logic: the Chinese in San Francisco, the Irish in Boston and Philadelphia, the Italians and Russians in New York City all fell under special hygienic scrutiny, in part because they personified alien racial status. They were, by their look and sound, displaced persons and, as such, under the precepts of social contagionism, inherently dangerous.

Commercial advertisement and sophisticated antituberculosis propaganda, which the progressive Jane Addams called "social advertisement," worked in tandem to threaten the uncertain professional monopoly of regular medicine even while popularizing its appeal.[102] The same institutional, social, and economic developments that produced and were produced by national mass marketing, advertising, and brand-naming were also at work in transforming the logic of public health from a highly located, particularistic approach to a standardized, generalizable one. Moreover, the new public health movement was itself a tremendous stimulus to the advertising, invention, and production of new hygienic products. The commercialized public health publicity was displacing medical authority, as one observer noted in 1911:

> True, as the physician has lost his monopoly on knowledge of health laws, the layman has found health stories and health work absorbing. True, our bill-boards, street-cars, magazines and newspapers afford innumerable evidences that huge factories have been built and stores started to cater to the new appreciation of health laws — e.g., vacuum cleaners, incinerators, sanitary drinking fountains, sanitary underwear, ventilated shoe-soles, disinfectants, "spotlesstown" soaps, health cereals, etc.[103]

Nevertheless, the author insisted, the physician could remain essential to the continued enlightenment of the lay public, if only he would embrace this educational mandate.

Despite the general reorientation of medicine toward public health education, such preventive social strategies popularized medical knowledge but could not replace medical research as the primary basis for professional authority. Public health workers became increasingly familiar with the techniques and importance of microscopy and bacteriology and relied increasingly upon the laboratory in making reports on morbidity and mortality.[104] This new technical sophistication within the emergent public health profession augmented danger: in addition to the conventional connections between contagion and visible filth, there emerged an increased, often exaggerated, appreciation for the dangers of invisible germs.[105] During the late Progressive years, just before World War I, commercial advertisers adopted the tenets of the new public health and strengthened their reliance on the scientific discourse of germ theory to promote particular cleanliness products.

In the decades before the development of antibiotics, infections were a serious threat in cases of even minor injury. The very possibility of preventing infection through antisepsis and asepsis made the threat of infection, long a fact of life, suddenly intolerable. Military officials began to distinguish between medical and military casualties, noting during the Civil War that they lost more soldiers to disease than to enemy attack.[106] The Reverend Dr. Bellows, the president of the U.S. Sanitary Commission, after a tour of western encampments, had stressed these invisible dangers:

> The perils of the actual battle-field are nothing to such men; the injury their open enemies can do them, almost not worth thinking of; but will malaria, fever, pestilence — irrational and viewless enemies — be as little dangerous? No! It is before these inglorious but deadly foes that our brave boys will flinch; before their unseen weapons that they will fall![107]

During both the Spanish-American War and the building of the Panama Canal, journalists portrayed specific diseases as greater threats to American imperial objectives than the political or military opposition. Yellow fever and malaria became the focus of intense research in part because of their military significance, and their "conquest" was identifiable with American expansion. These diseases, inasmuch as they were spread by insect carriers, helped reify the invisible world of bacteriology.[108] They also added specificity to the historical connection between hygiene and the military.

During the early years of the Great War in Europe, before the United States became engaged militarily, distant dangers of the battlefield entered American homes through advertisements in popular magazines. In a full-page, color advertisement in the *Ladies Home Journal,* Johnson and Johnson Company advertised their Red Cross absorbent cotton, a product developed in Europe for war use:

> There are invisible, ever-present living particles (called Germs) everywhere. They quickly lodge in the open flesh by contact with the air, dirt, unboiled water, clothing, skin, unclean bandages, and unsterilized hands. The consequences may be blood poisoning, inflammation, gangrene, fever, lockjaw, gatherings and a train of complications.[109]

Just as the law enforcement metaphor in health propaganda arose from the actual involvement of police in health enforcement, the wartime medical crisis fostered a resurgence of a newly specified military metaphor in hygienic advertisement. This advertisement, printed two months before the United States declared war on Germany and evoking a distant but very real threat of war, encouraged the reader to believe in invisible, ever-present "Germs."[110] The notion of an invisible enemy reflected a singular aspect of modern trench warfare: the deadly opponent was always out of sight. Historians of the Great War report that combatants sensed this invisible danger yet complained, "there is nothing out there, nothing to contend against."[111]

As the fighting in Europe intensified, so did the war at home against invisible enemies, and American manufacturers of patent disinfectants were on the metaphorical front lines. Again, figurative language was not spontaneously generated in the minds of health propagandists; rather, military and criminological elements of hygienic discourse were in place prior to the massive health crisis that followed the war and were available both as ways of speaking of disease and as ways of speaking of war. Accordingly, the concentration of military manpower during the war and American soldiers' engagement overseas intensified public health concerns over venereal disease. As historian Allan Brandt shows in detail, syphilis became a "third column" against the Allied forces, a "more deadly enemy than the Hun." The militarization of syphilis in the context of the war reinfused with urgency the agonistic equation of war and disease.[112]

The 1916 poliomyelitis epidemic in the United States, and the worldwide pandemic of influenza that struck the United States in September 1918 and killed millions before ending a year later, made the military overtones of hygienic propaganda all too vivid, reinforcing concern with invisible but nonetheless powerful dangers.[113] No manufacturer flogged the invisibility theme harder than did the New York–based Lehn and Fink Company, producers of Lysol all-purpose disinfectant. "Disease Besieges Every Home," women read in the *Ladies Home Journal* in the summer of 1917: "From every case of sickness hordes of invisible disease germs swarm forth to spread contagion." The threat was "in the very air," which, "laden with disease," created enemy agents of one's own family. "You . . . may carry germs in the folds of your clothing to your home." Insects, summer breezes, even "the romping of children" all brought the "ravages of health into the home." The very act of housecleaning nourished dangerous germs, stirred them into the air, to besiege especially "the cleanest appearing homes," "where ordinary soap and water were frequently employed."[114] Noting that "Army disinfection is compulsory," makers of Lysol urged the readers of the *Ladies Home Journal* to do their "share" to "offset the baleful effects of . . . ignorance" among those who "refuse to believe in germs — and their danger — because they can't see them." Disinfection had changed Panama from a "pest hole" to a "health resort"; the enlightened housewife could turn her "haunted house" into a "germ-proof home."[115]

Again, the elaboration of the political metaphor was double-edged: as politicians like Theodore Roosevelt touted the vision of American military power as the world's police force, the notion of hygienic "police" merged with a more warlike vernacular in health literature, both commercial and professional. Lehn and Fink Company marketed Lysol disinfectant as the "guardian" of home life against the "invisible menace that threatens every home, all the time." "Millions of deadly germs will, in a few hours, grow in a garbage can, an unwashed milk bottle, a clogged kitchen sink, or anywhere that flies gather and breed."[116] Graphic, detailed advertising copy, similar in cadence and vocabulary to Charles Chapin's evocative prose, made tangible the invisible danger of germ invasion. Makers of Lysol flattered their customers, whose very patronage placed them in the avant-garde of preventive medicine. "The baleful activity of the ignorant — the chief cause of the spread of disease — necessitates constant vigilance on your part." This argument implied that germs are insubstantial to those who are either ignorant or unfit for their offices. "If germs were as big as rabbits," one advertisement read, "they would soon cease to menace mankind."

Like the tailors of the emperor's new clothes, advertising copywriters must have found the invisibility of microbial life both a nuisance and a boon. Being invisible to the naked eye, germs could be conjured everywhere, but being invisible, germs were also easily ignored by an ostensibly unenlightened public. Advertisers vacillated between emphasizing the invisibility of these invaders and personifying them. In the "invisible" strategy, the effort was to promote appreciation for the unseen dangers of germ life, respect for scientific authority, and faith in the (equally invisible) benefits of a given commodity. For example, Lever Brothers promoted Lifebuoy soap as the "sensible soap" that "destroys the germs you can't see by removing the dirt you can see.[117] Advertisers could not, and did not, however, rely entirely on the invisible to promote hygiene products. Illustrators also adopted the opposite tack, of personifying germs, making them, in effect, "as big as rabbits" — and a lot more threatening. Illustrators personified germs not as clean-cut, law-abiding, Anglo-Protestants but as members of contemporary out-groups: as African Americans, Jews, Italian Americans, criminals, vagrants. Again, the analogy cut both ways: if immigrants, African Americans, and Jews looked suspiciously dirty, diseased, and criminal, disease germs, under microscopic examination, looked suspiciously foreign, dark-skinned, Jewish. Enhanced by these metaphorical associations, nativist sentiments reached new depths after World War I.

In different local contexts, different human scapegoats were made to assume the parasitical role. A 1914 poster, "The Cold That Hangs On," issued by the Health Publicity Service, pictured a sight familiar to urban Californians and travelers in Asia: a man being carried piggyback. In cartoonist H. F. Thode's rendering, however, the burden bearer was a well-dressed Caucasian man, coughing and weighed down by a large Chinese man whose coat bore the label "A. Hangon Cold" but whose coat lining revealed the label "tuberculosis." This smiling figure warned, "They call me 'Only a Cold,' they do not see the lining; That's my chance to get a firm hold, while my host goes on declining." A caption warned, "A cold that 'hangs on' very often has a tubercular lining. Consumption discovered *early* is cured easily. If you have a bad cold that cannot be 'thrown off' find out the cause."[118]

The concurrent "crusade" for the euphemistic cause of "social hygiene" gave a sexual charge to this germinal discourse.[119] Women, in their ascribed childlike foolishness and their attachment to frivolous fashion, were a favorite target for male physicians' approbation. In 1912, *Munsey's Magazine* carried a long article entitled "How to Make Yourself Germ-Proof," which asserted, "The fact that more girls and women do not succumb to germ diseases demonstrates nature's effective provision for the self-killing of germs." Dr. William Lee Howard explained,

> There is scarcely a woman or girl who does not daily carry deadly germs to her lips and mouth. Dirty money, bills or silver, hat-pins, a strand of some dead Chinaman's hair, theater tickets, newspapers, programs, combs—anything and everything that she may wish to retain for the moment. It looks to me as if women never outgrew the baby age—everything they take hold of goes into their mouths.[120]

As accomplices to germ assaults, "Women will never be safe from germ diseases, from the simplest to the most horrible, until they keep their mouths for eating only—and, of course, for conversation." The doctor was particularly appalled by the fashionable uses of human hair: "They will never be free from the danger of skin ailments and baldness—all germ diseases—until they stop putting the hair of dead men upon their scalps." He deplored the various armatures designed to support popular hairstyles: "Wire cages, rat-traps, and other cannibalistic head gear. . . . Even a woman's own dead hair is an abomination and an enemy."[121] The woman as accomplice to her own destruction and enemy to her own body, in league with ostensibly uncivilized alien men, in this charge stood as a metonym for graver transgressions of a sexual and economic nature. The notion of "freedom" from disease also played subtly to the prevailing logic of so-called white slavery, itself a complex political language that, by an inverted historical analogy, figured white women as sexual slaves to black, Jewish, Catholic, or Chinese men.[122]

Social contexts inspired specific imaginary villains. These images were not limited, however, to special local political struggles but appeared in parallel forms across the country. Atlanta's black washerwoman, "Soap Sally," the Chinese laundryman, "A. Hangon Cold," the Irish "Typhoid Mary" and her counterpart, "Tuberculosis Bridget," and the generic ne'er-do-well "Jimmy Germ" not only personified racist and nativist sentiment but also located it geographically, culturally, and immediately within the familiar contexts of everyday domestic and (especially) urban life. The nation's most prestigious popular health magazine, the American Medical Association's *Hygeia*, contributed its share to the popular image of the germ as an alien criminal element. The "trouble maker" Jimmy Germ was a recurring character in *Hygeia*'s cartoons. A stubbled, skinny, hook-nosed creature, Jimmy Germ bore a striking resemblance to traditional caricatures of Jews and to contemporary caricatures of southern European immigrants and "the criminal type." Jimmy took perverse pleasure in victimizing those foolish, dark-haired children who neglected to wash and to brush their teeth. This archcriminal could be thwarted by the fair-haired child who "obeyed the laws" of health.[123]

Comical as this figure may seem, Jimmy Germ personified the notion of health and disease prevalent throughout this period. Moreover, this caricature and the story it represented was eminently consistent with a powerful train of thought connecting criminality to contagion and connecting both to immigration and subversion. As one medical officer at Ellis Island put it, "Insanity and mental defectiveness [among immigrants] are of grave concern from the standpoint of public health. The individual victim is predisposed toward crime."[124] For eugenicists in favor of immigration restriction, immigrants carried dangerous germ plasm as well as deadly germs; they harbored bodily parasites and became social parasites. Jimmy Germ reflected and reinforced popular anti-Semitic and anti-immigration sentiment: he appeared in 1923, a year of intense debate over what would become the 1924 Immigration Restriction Act.

The twenties had begun with the infamous Palmer raids, so named after Attorney General A. Mitchell Palmer, whose Bureau of Investigation rounded up four thousand resident aliens in more than thirty-three cities on suspicion of being afflicted with "a disease of evil thinking."[125] Palmer's medical figure of speech was very much in keeping with that of other immigration officials, some of whom viewed immigration restriction as a eugenic policy.

In the spring of 1920, Harry Laughlin, an associate of Charles Davenport, became an "expert eugenical agent" for the congressional Committee on Immigration and Naturalization. Two and a half years later, Laughlin issued his report on inmates of state institutions who were of foreign extraction. The report included photographs of these inmates, entitled, "Carriers of the Germ Plasm of the Future American Population."[126] Through photographic, statistical, and linguistic usage, such sociological studies reinforced cultural associations among criminality, feeblemindedness, disease, and immigration. As a function in part of eugenical discourse, crime too became medicalized in the late nineteenth century as new professional social scientists contested dominion over particular social problems.[127] The powerfully negative import of crime became increasingly associated with immigration: the problems of disease with immigration, and the problems of crime with disease.

In the social-contagionist idiom of American eugenics, certain public health innovations, such as Wisconsin's 1912 legislation banning common drinking cups at public water pumps, provided legitimizing medical analogies for later and potentially more controversial eugenical legislation.[128] Such usage expanded popular understanding of germ theory, confused popular medical vocabulary, and further strengthened cultural linkages among the concepts of crime, disease, and immigration. Contemporary eugenical discourse answered the problems of criminology with the borrowed authority of medicine; by its concern with matters of reproduction, contraception, and prophylaxis, it euphemistically addressed the more intimate problem of "social hygiene," specifically the very serious social threat of venereal disease.[129] In parallel fashion, contemporary commercial advertisements for hygienic products euphemistically addressed on a more personal scale the problems of "feminine hygiene." The language used by advertisers to promote some disinfectant products was a language of double entendre, which capitalized on the term *hygiene* as a synonym for contraception and on the homologous relation between the bacteriological usage of the word *germ* and the eugenical and biological usage of the term *germ plasm.* Thus the makers of Lysol recommended its use, until the 1950s, for "personal hygiene," that is, as a contraceptive and an abortifacient. There were several documented deaths arising in the 1930s from such use.[130]

In subtler ways, advertisers also played upon the particular anxieties of women alone, or of women as the guardians of female children, about personal crime. A 1917 wartime advertisement for Listerine antiseptic pictured a woman and her daughter alone, dressed for travel, in a featureless room. The child is reassuring: "Here it is, Mother," indicating a Listerine bottle in their suitcase. The slogan reads, "Listerine—the *safe* antiseptic."[131] Women's greater mobility and increased commercial involvement—in some measure forced upon them by war conditions—created a new opportunity for advertising copywriters. Johnson and Johnson's antiseptic soap was the answer for "mothers, necessarily mingling with every type of humanity in the stores, theaters and at social affairs," who wished to avoid "carrying infection home to their children."[132] A decade later, Lysol was represented as a German shepherd dog standing guard over a little girl. Consumers were warned, "She seems safe, during the day, playing her games with her big dog to watch over her. And she seems safe at night, asleep in her clean little bed." Evoking a mistrust for appearances common to early Depression-era advertising, the copywriters cautioned: "But is she really protected? No matter how carefully you clean your house with soap and water, her worst enemies are still likely to be left as active, as deadly as ever. Soap and water cannot safeguard your home against the germs which carry the diseases most dangerous to little lives."[133]

The language of criminology, protection, and danger expanded into other realms of personal hygiene associated not with disease but with sexuality. Advertisements for Bristol-Meyers' Mum deodorant are particularly interesting in that they simultaneously and explicitly portrayed the consumer as victim and criminal. In a 1937 advertisement in the *Ladies Home Journal*, a fortune-teller warns a woman, "You have an enemy—a beautiful blonde—IT'S YOURSELF!" The woman does not realize it is "her own failings that defeat her," foolishly believing that her "daily bath protects her." The smartest women choose "one unfailing way to be safe at *all* times: The daily Mum habit!"[134] The copy strongly implied "protection" not only against underarm odor but also against all of the risks of intimacy. Women could avoid the "danger," even "disaster," of becoming "underarm victims" by using this product.[135]

Although germs and disease were not usually explicitly included in the body odor drama, the dramatic framework of danger, risk, victimization, and protection that characterized popular germ theory was easily borrowed in service of deodorant sales. This "danger" theme also recalled much older beliefs about etiology and sexual reproduction: that odors alone were capable of causing both disease and pregnancy.[136] By the mid-nineteenth century, physicians could dismiss these beliefs as obsolete: "The old idea that it was only the *odor* or 'aura' of the semen that ascended into the female organs and impregnated the ovum, is too unfounded and obviously incorrect to need refutation."[137] Nonetheless, just as older etiological concepts of putrefaction, miasmas, and filth persisted in the popular culture well into the twentieth century, the connection between odor and sexual danger remained powerful enough to evoke, or to be expected by advertisers to evoke, both anxiety and a purchase.

The commercial language of danger from unseen enemies—criminals, foreign invaders, or both—that characterized popular representations of germ theory thus carried into neighboring hygienic domains. Advertisers continued to view these appeals as effective after World War II. A 1952 photograph of a lonely house at night, safeguarded only by an enormous bottle of Lysol, reinforced the threat of disease as crime. Nonetheless, copywriters anticipated skepticism: "You cannot *see* a disinfectant kill germs; you must buy it on faith. Pin your faith, therefore, to the product that is endorsed by those who insist upon seeing before they believe. Physicians everywhere urgently recommend Lysol. Follow the lead of those who know: use Lysol for personal hygiene, and for home disinfection." Although the commercial uses of fear, including fear of crime, are not in themselves remarkable, we should remember that these advertisements, and hundreds like them, were not promoting security alarms, ballot initiatives, or handguns but household disinfectant, mouthwash, and deodorant.[138]

Marketing Fear

The conception of the germ as a deadly enemy, a sly criminal with malicious intent, or a dangerous foreign invader was in part the result of efforts on the part of educators, advertisers, and public health officials to make real the invisible threat of microbial life. This concept of disease played upon the other anxieties attendant on urban living in the early twentieth century: fear of personal and violent crime, of immigrants, of strangers in general, of moral pollution and cultural loss. It also created images of health that were highly militarized.

The appearance of policemen, guard dogs, military language, and criminological terms all contributed to the talismanic power of hygienic and disinfectant products. These symbols seemed to suggest that the value of Lysol, Listerine, and Red Cross absorbent cotton went far beyond the merely hygienic, to provide other forms of protection against the many dangers of modern life. Moreover, by placing health and disease in a moral and legal context, these words and images underscored the "new" public health philosophy of personal responsibility through intelligent participation in the market.[139] By the 1930s, "health sinners" had become "health criminals," and one could act as accomplice to the criminal germ against one's own body.[140] The uneducated or uncooperative consumer became "her own worst enemy."

The 1930s saw health reformers and advertisers continuing to elaborate both the criminological metaphor of disease and the law enforcement logic of hygiene. In keeping with contemporary pedagogical philosophy, in which drama played a major role, the American Medical Association publication *Hygeia* was filled with songs, cartoons, playscripts, and poems designed for classroom use. In the 1930 poem, "A Syncopated Health Trial," the presiding judge faced the accused sufferer of tooth decay: "You say this poor, pale lady is guilty of neglect, Has failed to use her toothbrush, until her health is wrecked?"[141] The accused had committed a crime against her own health, but under the assumptions of the new public health, this was also a crime against society. Even children could become health criminals. Advertisements for Kleenex tissues portrayed microscopic views of Kleenex fibers that retain "germs," preventing their spread; the accompanying copy warned mother, "Don't let your child menace a schoolroom" with an old-fashioned handkerchief. "Germ-filled handkerchiefs are a menace to society!"[142] Another writer for the AMA publication characterized a child's hand as a "living culture plate," which led him to "indict" thumb-sucking on a "sanitary charge."[143]

Although the crime metaphor remained a central organizing concept behind popular epidemiological culture, the specific kinds of crime associated with disease changed with the times. Newspaper headlines of mob activities during the 1930s were easily translated into the advertising vernacular: an enormous housefly looms above the caption: "Help disarm this gangster!" The copy explains, "You lock your door securely against criminals . . . but perhaps lock in a far more dangerous enemy . . . germs of infection! Do YOUR part in the never-ending war against Infection. . . . Use Lysol."[144] In this particular advertisement, the imagery of crime, criminals (gangsters in particular), enemies, and war converge, so that the housewife is bound not only by motherly duty but also by patriotism to purchase Lysol for the protection of her person, home, and country. Although it is somewhat unusual to see all of these themes working in a single advertisement, the net effect of hygienic advertising in general was to confound these categories, through association, and to make them almost interchangeable. The uniformed protection officer, whether of the civil or military police, was the ubiquitous symbol of proper hygienic law, well obeyed or well enforced. Although the advertising industry and health popularizers gradually abandoned the scare tactics of the Depression years, most such writers and illustrators retained the general criminological framework, complemented by military analogy, in which hygienic propaganda had been cast since before the turn of the century.

Some of these later images are perplexing when viewed apart from their historic-semantic context. A full-page advertisement in a 1943 Life magazine, for example, pictures a photograph of seven articles: a bottle of Listerine, a stethoscope, a drinking glass, a thermometer, a nurse's cap, a roll of sterile bandages, and—a police officer's hat. "For Life's Little Emergencies," reads the caption, but the police hat gives statal force, and historical resonance, to the quasi-medical authority of the other items pictured. The symbol of police power worked in two ways at once. It served to reinforce the notion of guaranteed protection against hygienic crimes perpetrated by others and to give notice of the serious consequences to those who might flout the advertiser's friendly advice. Simultaneously, it served as threat and reassurance, backing commercial advertisement with the emblem of coercive state power.

The blue hat, seemingly out of place among the paraphernalia of first-aid nursing, belongs to a nineteenth-century tradition of hygienic representation and public health practice. It purchased masculine authority and protection for the feminine consumer. As part of a vast, intricate, historical system of political semantics, it figuratively bound the police power of the state to the healing power of medicine, through the trivial inventions of commerce, at once furthering and denaturing the militaristic-cultural equation that permits such barbarous euphemisms as "surgical strikes," the "war on drugs," and "ethnic cleansing."

CHAPTER FOUR

"Just Say No"

Risk, Behavior, and Disease
in Twentieth-Century America

Allan M. Brandt

Wᴴᴼ ɢᴇᴛs sɪᴄᴋ? And why? These are among the most profound human questions. Every society has attempted to develop some way to account for the phenomena of disease and death. What are the causes of mortality? And when, in the course of life, does death come? In part, these questions reveal deep moral and philosophical ideals, but they also engage other powerful cultural values and beliefs, as well as specific social, medical, and scientific investigation. The nature of morbidity and mortality have historically been subject to dramatic shifts, even over relatively short periods of time.[1]

In the course of the nineteenth century, patterns of health and disease in the United States underwent a radical transformation. This change is reflected in major demographic indicators, the nature of the practice of medicine, as well as cultural norms and values concerning health and disease. Not only do Americans live longer (life expectancy rose from forty-seven years in 1900 to almost seventy-five in 1988), they now die from different causes.[2] Although these changes have been widely noted and often celebrated, their larger implications for the ways Americans think about health and disease have gone largely uncharted. Nevertheless, as this chapter suggests, changing perceptions about what causes disease—the nature of risk, behavior, and responsibility—reflect powerful cultural beliefs. And, in turn, these beliefs affect, both implicitly and explicitly, patterns of social behavior and the organization and delivery of health care. The chapter examines, in somewhat speculative fashion, the cultural implications and meanings of this "health transition" and explores how theories of disease causality may reflect powerful social and political ideologies concerning risk and responsibility.[3]

At the turn of the twentieth century, infectious diseases constituted the predominant causes of death. The widespread occurrence of serious epidemics of malaria, smallpox, yellow fever, and cholera throughout the eighteenth and nineteenth centuries had created a shared sense of vulnerability.[4] Epidemics, with their sudden onset and often disastrous impact, created a perception of a world clearly not under human control. Efforts to arrest these contagions through sanitation, quarantine, or medical intervention often failed to have any influence on the course of disease.[5] As late as 1918, an epidemic of Spanish influenza underscored the devastation that disease could wreak in a short time. Worldwide, flu claimed the lives of perhaps as many as twenty million people (550,000 in the United States).[6] If epidemic disease did not result in fatalism, it did at least suggest the limits of human intervention to alter fundamental biological processes. Although explanatory models of disease before the late nineteenth century often centered attention on religious meanings and responses, this reflected in part a deeper recognition of human limits in the face of severe biological constraints.

Epidemics merely represent the fluctuation in more basic patterns of endemic infections. Tuberculosis and pneumonia actually constituted a more serious risk to life than did the intense, but time-limited, epidemics. Infant mortality (deaths during the first year of life), for example, which remained prevalent into the twentieth century, was typically caused by common enteric infections of newborns causing diarrheal disease. As late as 1900, infant mortality in the United States remained at more than a hundred per one thousand live births. By the mid-1930s, the prevalence of infant mortality had fallen precipitously to approximately sixty per thousand.[7]

Recognizing that diseases were selective in their targets—some suffered and died while others, socially and geographically proximate, were spared—led to medical theories that emphasized the hereditary quality of susceptibility to particular diseases.[8] Physicians often emphasized the importance of an individual's "constitution" or "diathesis." For most of the nineteenth century, for example, physicians and public health officials emphasized the nature of familial vulnerability to tuberculosis, the most prevalent disease causing death. While physicians and social critics debated the relationship of moral turpitude and disease, the nature of the worthy and the unworthy poor, both epidemic and endemic disease were reminders of the precariousness of life.

BY THE LATE nineteenth century, the development of the germ theory of disease had radically altered both medical and social meanings of disease. The preeminent axioms of the germ theory were Robert Koch's "postulates" — the rules to live by in modern biological science. Koch, after discovering the tubercle bacillus in 1882, had concluded that a single pathogen was invariably associated with a specific disease; that the organism could be isolated from a lesion, grown in a pure culture, then used to reproduce the disease. The power of these postulates, and their experimental elegance, transformed the biological sciences. The search for disease-causing microorganisms began in earnest; by 1900 nearly thirty "causal" organisms had been definitively identified under microscopic scrutiny.[9]

Soon after the discovery of a number of these organisms, researchers came to understand that it was possible for individuals to be infected but to remain free of disease. This paradox of the "healthy carrier" undermined the conceptual clarity of the germ theory. It soon became clear that the pathogenic agent was a necessary, but not a sufficient, cause of disease. In fact, this was particularly true with tuberculosis. Researchers found that in many urban areas a majority of inhabitants were infected but remained healthy. As Rene Dubos later pointed out, "Infection is the rule and disease is the exception." While the notion of specific etiology, the very basis of the "biomedical model," became the focus of medical investigation and treatment, according to Dubos, "few are the cases in which it has provided a complete account of the causation of disease."[10] Other important anomalies — instances where, for one reason or another, scientists failed to satisfy the rigor of the postulates in spite of the presence of an organism and its association with a disease — also came to be recognized. With medical science's focus on the organism as the cause of disease, the significance of the social environment was diminished. Other critics suggested that medical science had begun to focus on the specific aspects of pathogenesis, losing a perspective on the "whole" patient.

Although biomedicine, and the germ theory in particular, led to a far more complete understanding of the causes and nature of many diseases, it did not lead to immediate or effective therapeutic modalities. Nevertheless, infectious diseases did decline, and in fact, as more demographic data were sifted, it eventually became clear that most important infectious diseases were declining even before the elucidation of the germ theory.[11] The leading causes of death in 1900 were tuberculosis, pneumonia, and diarrhea. In 1990, heart disease, cancer, and stroke constituted the principal causes of death; cancer and heart disease alone accounted for almost 60 percent of all deaths.[12] What accounts for this truly revolutionary change in the patterns of disease? Although many associated this epidemiological and demographic shift with the rise of the germ theory and "scientific" medicine, most data suggest that this transformation was well under way before medicine developed any decisive technologies to fundamentally alter patterns of disease and death. Changes in the material conditions of life — sanitation, nutrition, and birth rates — led to fundamental changes in patterns of infection and longevity.[13]

For any understanding of the relationship of culture and science, the problem of causation is critically important because it reflects directly on the question of responsibility for disease. Despite the anomalies implicit in the germ theory and the reductionist qualities of the biomedical model, the bacteriological revolution had the effect of "depersonalizing" disease. Under the microscope, diseases could no longer possess the same moral valence they had possessed in the past. In the increasingly secularized and rational world of medicine and science, microorganisms came to be viewed, almost unilaterally, as the cause of disease. This offered at least the possibility of disconnecting disease from its historic associations with sin, moral turpitude, and idleness.[14] There were, of course, a number of diseases that continued to have powerful moral meanings: mental illness, alcoholism, and the sexually transmitted infections, to name but a few.[15] But many diseases came to be seen as the result of a random chain of events that brought together a microorganism (a vector) and human beings. Disease was no longer seen as necessarily reflecting the personal attributes of the sick individual. The biomedical model had the effect of depersonalizing and secularizing disease.

It was in this historically specific context that the attribution of disease had the effect of reducing individual responsibility. Disease, in this biomedical paradigm, became a secular and scientific phenomenon, freed from traditional moral linkages. This accounts for the eagerness with which the founders of Alcoholics Anonymous, for example, seized the notion that alcohol dependence was a disease.[16] It was not so much a desire to medicalize the phenomenon of habitual alcohol consumption as it was to free those with the habit from moral stigma, to "remoralize" the behavior.[17] A similar phenomenon occurred in the early twentieth-century movement to transform homosexuality from a crime into a disease, an indication of psychopathology. The implication was that those who suffered from the "disease" should be relieved from the traditional culpability associated with their behaviors.[18] Disease implied a lack of volition or, at least, a failure of individual agency. The concurrent move to expand the nature of the insanity defense during the first decades of the twentieth century is yet another example of this significant trend.[19]

In the biomedical model, the occurrence of disease largely came to be seen as the result of discrete phenomena. Susceptibility was defined as a lack of antibodies to a particular organism; resistance could be acquired through natural exposure to the organism or through vaccination to stimulate antibody production. In keeping with this particular paradigm, biomedicine focused on destroying pathogens or inducing resistance to them. With the introduction of sulfa drugs in the 1930s and antibiotics in the 1940s, the germ theory had spawned effective technologies to combat infection. The promise of Paul Ehrlich's "magic bullets" — specific chemotherapies used to root out and destroy "invading organisms" — had at last been realized. Diseases that a mere decade earlier had posed a serious threat to life could now be quickly and definitively treated; antibiotics now routinely saved those previously damned. The golden age of American medicine had begun.[20]

In spite of the critics and theoretical limitations of the biomedical model, it is not difficult to recognize its appeal. If disease was caused by a single microorganism, then destroying these microorganisms — either in the environment (through sanitation) or in the body (through therapy) — offered the promise of conquering disease through definitive technologies. The elegance of magic bullet medicine has remained one of the most compelling metaphors in modern medicine.[21] The possibility of dramatic cures after centuries of stalemate inspired a new awe regarding scientific investigation and medical intervention.[22]

THE RISE IN life expectancy, the end of epidemics of infectious disease, and the growth of effective and dramatic medical interventions all led to an era of rising status and authority for the medical profession. Medicine came to be distinguished from biomedicine. Medicine was the undifferentiated past, in which doctors could offer patients little beyond support and theory; biomedicine was the result of a revolution in bacteriological and immunological science. The "miracles" of modern medicine, dispensed in doctors' offices as well as in technologically sophisticated hospitals, had brought new respect and acclaim to a profession that only a generation earlier had been little more than a competitive trade.[23] In the nineteenth century, as all forms of regulation eroded, it become increasingly difficult to distinguish among a wide variety of medical sects, local healers, quacks, and so-called regular physicians, who possessed a specialized education. But by the early twentieth century the profession was able to consolidate its authority and power, in large measure as a result of the rise of the germ theory and increasingly esoteric, privileged knowledge of science.[24]

During the course of the twentieth century, the epidemiological shift to the predominance of noncommunicable, chronic diseases exposed the problems inherent in magic bullet medicine. For the persistent problems of cancer, heart disease, and stroke, diseases in which it was impossible to specify a single cause, it proved impossible to specify a single solution. And even for infectious disease, specific therapies proved to have limitations—organisms could become resistant to previously effective chemotherapies. Even when effective treatments were available, they could not always be delivered on a timely basis. Moreover, the notion of specific causality neglected the complex interactions between agent, host, and vector.[25] Clearly, some human beings were more vulnerable than others. As a result of heredity, geography, or environment, some individuals were apparently more susceptible to a number of microorganisms.

Implicit in this shift from infectious to chronic, systemic disease was a transformation in the meaning of disease and the assumptions concerning its causes. Epidemic infectious disease was typically perceived to be the result of external forces. During the nineteenth century, physicians had debated whether disease resulted from environmental decay—miasmas, poverty, pollution—or from foreigners bringing contagion to new soil; disease was a "visitation."[26] Disease was episodic, and it did not originate or reside in the body. The chronic diseases of the late twentieth century were preeminently diseases of the body. Diseases were no longer caught, they were acquired.[27] The limits of the germ theory to address systemic chronic disease led to a new recognition of environmental and behavioral forces as determinants of disease in the postwar era.

This change in patterns of disease ultimately challenged the ascendancy of the germ theory. By the 1960s, the promise of the biomedical revolution had begun to fade. Despite the firepower of magic bullets, these drugs simply could not target the systemic chronic diseases, which now accounted for the preponderance of deaths in the United States and other developed countries. The war on cancer, a national commitment to finding a cure for cancer in the laboratory, had stalled as the promise of any definitive victory dissipated.[28] The technological "fix" for the remaining problems of disease proved illusory, despite remarkable technological advances. Cure proved the exception in the battle against the chronic diseases of later life.

New epidemiological studies, recognizing these dramatic changes in patterns of disease, began to reevaluate the basic medical principles of causality. As a result of the changing scientific emphasis in identifying the causal agents of diseases during the late nineteenth century, the discipline of epidemiology had undergone fundamental change. During the early nineteenth century, public health reformers like Edwin Chadwick in Great Britain, Lemuel Shattuck in the United States, and Louis Rene Villermé in France collected voluminous social data in communities in an effort to identify those factors—ranging from housing and work to pollution and poverty—that might account for particular patterns of disease prevalence.[29] After the elucidation of the germ theory, the emphasis in public health came to center on finding and destroying (wherever possible) the noxious microbes. With the bacteriological revolution, interest and concern about general social conditions as a cause of disease declined, as epidemiologists turned their attention to laboratory diagnosis.[30] Statistical inference seemed a weak tool in comparison to the sophisticated biochemistry and microscopy of the laboratory.

The significant change in patterns of disease that occurred during the first half of the twentieth century, however, encouraged a "new" epidemiology. From tracking microbes, which were uniformly seen as the "cause" of disease, researchers began to identify *risks:* the social, environmental, and behavioral variables that were statistically associated with patterns of chronic disease.[31] The nature of causal inference in the sciences became a contested domain in the context of this demographic transition. Although some researchers continued to explore theories of specific causes of chronic disease, especially hereditary predispositions, the recognition that many factors were likely to contribute to the development of diseases such as cancer and heart disease led to a revolution in epidemiological technique. The rise of modern biostatistics and the development of controlled prospective trials offered the opportunity to explore the relationship of a host of environmental and behavioral variables to patterns of health and disease.

A PRIME example of this complex historical process of legitimating new approaches to causal inference lies in the history of the cigarette. The recognition that cigarette smoking causes serious disease followed by the decline in smoking is characteristic of the postwar shift regarding risk, disease, and behavior. It was no simple task to "prove" that cigarettes cause disease. By the end of World War II, cigarette smoking had become an enormously popular social behavior in American life, an icon of the consumer culture. The cigarette, an unusual and somewhat stigmatized form of tobacco consumption at the turn of the twentieth century, had become—through creative marketing and industrial consolidation—a symbol of affluence, leisure, personal power, and attractiveness. But despite its categoric success, concerns persisted about its impact on health. Lung cancer, virtually unknown in 1900, had begun to rise alarmingly.[32] By the 1940s, spurred by the findings of life insurance actuarial studies, major prospective studies of cigarettes, lung cancer, and mortality began in the United States and Great Britain. These studies concluded that cigarettes constituted an enormous risk to health, not only as a cause of lung cancer and other lung diseases but also as a contributing factor in heart disease.[33]

The epidemiological studies of cigarette smoking culminated in the surgeon general's report of 1964. The report, written by a group of eminent scientists under the auspices of the surgeon general, reviewed the epidemiological
evidence indicting cigarettes as a cause of disease and concluded that these
studies, in fact, conclusively demonstrated the risks of smoking. Implicit in this
research was an influential critique of the whole notion of specific causality.[34]
This type of quantitative epidemiology touched off an important debate within
the scientific community about the nature of causality, proof, and risk. At stake
were the very epistemological foundations of scientific knowledge: How do we
know what we know? What is the reliability of causal inference from statistical
data? At the basis of the epidemiological argument was the clear limitation of
laboratory experimentation for making determinations about probability and
risk. The debate about smoking and health revealed an intraprofessional battle
between epidemiology and laboratory science (the science of germ theory) —
their values, assumptions, and expectations.

Pressured by a growing consumer movement concerned about the dangers
of unregulated products and by the voluntary health agencies eager to offer the
public an approach to controlling chronic disease, the surgeon general finally
acceded to studying the problem of the relationship of cigarettes to disease.
The surgeon general's report made a fundamental contribution to medical
studies of causality. Members of the committee realized the complexity of the
relationship — of the impossibility of saying simply that smoking causes cancer. Many individuals could smoke heavily throughout their lives and apparently never suffer adverse consequences; cause implied a process in which *A*
would, by necessity, lead to *B*. Therefore, they acknowledged the complexity:
"It should be said at once," the report explained, "that no member of this
committee used the word 'cause' in an absolute sense in the area of this study.
Although various disciplines and fields of scientific knowledge were represented among the membership, all members shared a common conception of
the multiple etiology of biological processes. No member was so naive as to insist upon the mono-etiology in pathological processes or in vital phenomena."
This statement, a clear criticism of the germ theory's emphasis on specific
causality, lent new credibility to the entire field of modern epidemiology. The
report underscored the nature of the social process of scientific proof. This authoritative study now constituted the "proof" that cigarettes "cause" cancer
and signaled the beginning of a major battle between the tobacco industry and
public health forces in the United States.

Congress responded to the surgeon general's report by legislatively man-
dating that cigarette packs be labeled. The Federal Cigarette Labeling and
Advertising Act of 1965 required that all cigarette packages carry a warning:

"Caution: Cigarette Smoking May be Hazardous to Your Health." Given that the surgeon general had found that smoking *causes* lung cancer, the warning was remarkably weak, indicating the effectiveness of the tobacco lobby on Capitol Hill. It further reflected the relative lack of experience most legislators had with scientific findings, especially if they were contested. At the hearings concerning this legislation, tobacco spokesmen challenged the findings of the surgeon general. By treating all perspectives as those of "interested" parties to be brokered in the political process, members of Congress sought compromise. Moreover, powerful economic interests, especially of tobacco-growing states, acted forcefully to moderate any regulatory initiatives.[35] Nevertheless, as scientific studies collected in subsequent surgeon general's reports continued to indict the cigarette as a major cause of serious disease, Congress took additional action. In 1971, the label was changed to "Warning: The Surgeon General Has Determined that Cigarette Smoking Is Dangerous to Your Health." And, in 1985, four rotating labels were mandated. Although originally viewed as an educational and regulatory measure, cigarette labeling served the purpose of shifting responsibility for smoking and its risks from the industry to the individual smoker. Cigarette smoking—given the clear warnings printed on every pack—had become the preeminent example of a "voluntary" health risk.[36]

SUBSEQUENT epidemiological studies would expand the question of risk and behavior far beyond the cigarette. Beginning in the early 1950s, researchers in Framingham, Massachusetts, began to collect longitudinal data on the relation of a whole host of variables, including work, diet, and exercise, on heart disease and longevity.[37] In Alameda County, California, epidemiologists created the Human Population Laboratory to assess the causality of morbidity and mortality within a community, to look at a full range of variables as they contributed to both disease and health. Researchers concluded that three meals a day, breakfast every day, no snacking, plenty of sleep, and moderation in alcohol consumption were the keys to good health and long life.[38] But just as the germ theory contained powerful cultural norms regarding the meaning and significance of disease so, too, did the new theories that emphasized risk factors. The health truisms of *Poor Richard's Almanac* had come home to roost with the growing authority of quantitative epidemiological science. These prescriptions had gained the imprimatur of modern quantitative science.

With this logic, and a growing body of supporting data, the meaning of disease in American culture underwent a radical shift. By the early 1970s an emerging critique of modern biomedicine and medical technology centered attention on the question of responsibility for disease and its prevention. Few could disagree that disease prevention and health promotion were laudable goals. According to critics such as John Knowles, president of the Rockefeller Foundation, American society had reached the point of diminishing returns from its heavy investment in medical high technology and tertiary care. The failure, claimed Knowles, was in prevention of disease. The goal of health and longevity rested firmly with individuals, who in the last decades had forfeited their health in an "orgy" of greed, avarice, and overeating, the "diseases of affluence." As Knowles concluded, "Most people do not worry about their health until they lose it. Uncertain attempts at healthy living may be thwarted by the temptations of a culture whose economy depends on high production and high consumption." Knowles called for a return to Puritan values of self-discipline and moral restraint. Eager to reduce the "dole" implicit in rising health expenditures, Knowles suggested that "the idea of a 'right' to health should be replaced by the idea of an individual moral obligation to preserve one's own health . . . a public duty if you will." [39] According to this perspective, control of the persistent health problems of the United States—the chronic diseases of middle and later age such as cancer, stroke, and heart disease—depended directly on individual behavior and habits. Individuals could no longer rely on public health interventions, on the medical profession, or on the health care delivery system to solve the problems of disease. Rather, the mantle of responsibility in the quest for health would now be carried on the shoulders (or the backs) of individuals.

The emphasis on individual responsibility for health, as well as the epidemiological studies that demonstrated the significance of specific behaviors such as cigarette smoking, diet, and exercise offered the possibility of control over one's health. No longer would disease be viewed as a random event; it would now be viewed as a failure to take appropriate precautions against publicly specified risks, a failure of individual control, a lack of self-discipline, an intrinsic moral failing. By the late 1970s, Knowles's views had entered the political mainstream. In 1978, Secretary of Health and Human Services Joseph Califano, eager to find a way to reduce growing health expenditures, explained: "We are killing ourselves by our own careless habits. . . . We are a long way from the kind of national commitment to good personal health habits that will be necessary to change drastically the statistics about chronic disease in America. . . . Americans can do more for their own health than any doctors, any machine or hospital, by adopting healthy lifestyles." [40]

The idea that individuals can and should exercise considerable control over their health, of course, has deep historical roots. Good, clean living formed the basis of religious as well as medical ideals; disease, throughout history, has often been viewed as the physical price of sinfulness. In the nineteenth century, health reformers emphasized the importance of natural laws in determining appropriate behaviors and good health. Health promoters such as Sylvester Graham and John Harvey Kellogg defined a new asceticism, a moral economy of prescribed behaviors that promised prosperity and longevity.[41] While often justified on scientific or medical grounds, the Victorian exercise movement reflected deeper moral and religious sensibilities in an increasingly secular society.[42]

Since World War II, but especially since the 1960s, the fitness movement departed from its historical focus on general well-being, taking on the trappings of specific prevention of disease. In particular, clinical and basic research led to concrete conclusions about diet and exercise and their relationship to heart disease, stroke, and cancer.[43] Epidemiological findings of major prospective studies were repeatedly cited to justify calls for radical alterations in diet, alcohol consumption, and exercise. Although the fitness revolution preceded many of these data, medical science seized the movement and made it its own. The first major medical studies, for example, on the relationship of exercise to cardiological function did not appear until the early 1970s, but the medical profession soon acted to legitimate the cultural appeal of vigorous exercise. The boom in physical exercise of the 1970s and 1980s constituted a powerful shift in cultural values as well as actual behaviors. According to most surveys, one-third of all Americans reported engaging in vigorous exercise at least three times a week by the late 1980s; more than three times as many as the early 1960s. As many as fifty million people reportedly jogged regularly, up from twelve million in 1976. Similar data were reported for swimming and aerobic exercise. An estimated twenty-four million women regularly attend aerobic workout classes. *Jane Fonda's Workout Book* became one of the best-sellers of 1976, with some 1.25 million copies sold; her exercise videos soon became among the most popular commodities in that medium.[44]

While some observers suggested that the fitness movement possessed an implicit critique of medical authority and intervention, in practice, organized medicine cooperated vigorously with the crusade for physical fitness. After all, two essential epidemiological markers of health—blood pressure and cholesterol—remained the province of the profession. The "stress test" combined exercise and the evaluation of cardiological function in the doctor's office. Those who exercised could measure their progress (or failures) only through these fairly sophisticated medical calibrations. In a new twist on the doctor-patient relationship, the doctor became a health "monitor." Exercise became yet another aspect of the medicalization of American culture.[45]

Not unlike exercise movements of the past, the modern fitness revolution reflects the desire to assert control, to reduce the risks of disease or premature death. Exercise promises mastery and self-efficacy. The commitment to exercise promises more than health — it promises personal redemption. The zen of running has been described by a number of its advocates (such as cardiologist George Sheehan) as offering "rebirth" and "salvation." According to Sheehan, running promised "self-betterment" and psychological strength.[46] The fitness revolution, notably, does not focus on competitive or team sport; the battle is within. The goal is victory over the uncertainties of the body. Running, aerobic exercise, and modifications of diet all require harsh exertion and self-denial. In this respect, fitness can be viewed as a critique of the consumer culture, which has eroded the values of discipline and denial. But the consumer culture has rediscovered these values and developed a vast array of commodities for fitness: shoes, machines, and health clubs. The effectiveness with which the consumer culture *captured* the fitness revolution is testament to its contemporary tenacity.

The movement for individual health has also had major effects on patterns of food and alcohol consumption. Data collected during the 1980s indicate significant shifts in attitudes and behavior regarding nutrition as more findings appeared specifying health risks of diets high in cholesterol and fats. A majority of Americans reported attempting to reduce fats in their diets; egg consumption declined, while consumption of fish, raw vegetables, and whole grains rose dramatically. Similar shifts in alcohol consumption also occurred. In 1985, according to the Gallup poll, 55 percent of those surveyed agreed that "using alcohol for enjoyment is generally a bad thing." Moreover, 38 percent contended that the use of alcohol is morally wrong. Between the mid-1970s and 1989, the portion of adults who consumed alcoholic beverages declined from more than 70 percent to 56 percent.[47]

It would be easy to view this shift from an ethic of behavior and health emphasizing moral rectitude to one freed from such moralism — good behavior could now be grounded in biomedical rationality. But to view this transformation in attitude and behavior in this way would obscure important aspects of the current ideology of health and fitness, which reflects deep cultural anxieties about not only health but also bodily appearance.[48] If the fitness revolution is driven by scientific findings about risk and behavior, it also has a powerful moral and prescriptive dimension. In this sense, the fitness crusade of the

1970s and 1980s bears striking continuities with the Victorian campaign of the late nineteenth century.

The intense preoccupation with bodily health, in this respect, is thus not unprecedented. At historical moments of rapid change and social disruption, of apparent unpredictability and widespread perceptions of chaos in social values, the assertion of individual control becomes a paramount social goal. Focusing on the body as the object of that control becomes imperative. Such was the case in late Victorian America; the exigencies of industrialization and urbanization, the massive changes invoked by the radical transformation of society, led individuals to turn inward in a search for order.[49] And such has been the case in contemporary American culture since the mid-1960s. A period of domestic and foreign violence and war and a decline in political legitimacy and professional authority led to a resurgence of demands for individual health. The general failure to define technological solutions to a range of social problems, including energy supply, the environment, and especially chronic disease, led to an emphasis on establishing individual control over the vagaries and uncertainties of life, an emphasis on control over the body.

THE IRONY is that the process of pathogenesis is so complex and overdetermined that discussion of *cause* necessarily becomes a socially constructed and often contested domain. The environment, behavior, the relationship of host and parasite, and inborn resistances and vulnerabilities all play significant roles in determining patterns of disease and health. But at particular historical moments one set of causes may well be more compelling than another, more susceptible to human manipulation. Although science has increasingly specified and differentiated disease etiologies—and in so doing has had some dramatic victories over particular diseases—the process inevitably goes forward. New diseases are revealed by the nature of biological and demographic change, by changes in the environment and culture, and by changes in science and medicine. Historians can only attempt to define the devices a culture employs to give disease particular meanings and significance. The calls for individual responsibility and reform of personal behavior draw on deep cultural values and social psychologies concerning the nature of disease. At stake in the process by which health risks come to be elucidated and defined in twentieth-century America are several critically important political and economic conflicts. In this respect, cultural norms and values are fundamentally related to political conflicts. Is there a "right" to health care or a "duty" to be healthy? How do we distinguish between assumed risk and imposed risk as they related to patterns of behavior and disease?[50] Indeed, these distinctions probably reflect specific (and changing) historical assumptions about the nature of human behavior rather than any empiric reality.

Again, the cigarette provides a telling example of the historical issue. Shall we consider smokers ignorant and stupid for maintaining an "unnecessary" behavior that has been defined as highly dangerous? Or shall we recognize the power of advertising and cultural conventions as well as the power of the biological and psychological factors in addiction, all of which constrain individual choice? During the twentieth century, Americans have rejected fatalistic explanatory models of disease and its causes.[51] Social values emphasize that individuals can and should exert fundamental control over their health through careful and rational avoidance of risks. As effective as such values may be in defining healthful behaviors, they represent an important political and cultural irony. According to this behavioral ethic, those who continue to take risks must be held accountable for the results; but this emphasis on individual responsibility may deny broader social responsibilities for health and disease. Simply identifying individual behavior as the primary vehicle for risk — even with substantial epidemiological data — negates the fact that behavior itself is, at times, beyond the scope of individual agency. Behavior is shaped by powerful currents — cultural, psychological, and biological — not all immediately within the control of the individual. Behaviors such as cigarette smoking are sociocultural phenomena and are not merely individual or, necessarily, rational.

The emphasis on personal responsibility for risk taking and disease has come at the very moment when cigarette smoking is increasingly stratified by education, social class, and race. In 1985, 35 percent of blacks smoked, compared to 29 percent of whites. For college graduates, the proportion of smokers fell from 28 percent in 1974 to 18 percent in 1985; for those without a college degree, the decrease during the same period was from 36 to 34 percent. Thus to stress individual accountability is to deny that some groups may be more susceptible to certain behavioral risks, that the behavior itself is not simply a matter of choice. Nevertheless, individuals who "take" — note the voluntaristic bias — such risks are considered ignorant, stupid, or self-destructive. Identifying disease as an essentially voluntary process demonstrates the cultural imperative for control. If disease can be avoided by carefully following a set of prescriptions regarding personal behavior, then individuals can take control over their bodies and, thereby, their lives. By this logic, the persistent (and growing) differentials in the burden of disease between blacks and whites, for example, become but an artifact of the vicissitudes of individual behaviors.[52]

The very nature of these causal frameworks for understanding disease has been to make complex phenomena seem simple. In this process, important cultural values are expressed. Take, for example, the well-known "Just Say No" antidrug campaign of the 1980s. Implicit in this educational campaign, whose legitimate goal was to invigorate individual assertion over peer pressures, was the notion that simple self-denial can solve a complex social problem. The pathology of drug abuse is shifted; rather than the problem reflecting certain external social conditions, it now resides in the individual. Drug use becomes preeminently an aspect of individual agency: Just Say No.

To evaluate the power of this set of cultural assumptions one need take but a cursory look at the first decade of the AIDS epidemic. AIDS, of course, contradicted some of the most basic assumptions of postwar medicine: the era of infectious epidemics was supposed to be over; ours was a time of chronic diseases that struck late in life, debilitating the growing ranks of the elderly. AIDS strikes the young, is communicable, and is caused by a deadly pathogenic organism. But despite these characteristics, AIDS has been placed strongly within the paradigm of responsibility. If one "merely" avoids the risk behaviors associated with the transmission of the virus — unprotected sexual intercourse and the sharing of intravenous drug paraphernalia — one can avoid AIDS. Therefore, infection is a clear, and usually terminal, marker of individual risk taking, of engaging in behaviors held to be deviant or criminal. According to this view, those who are infected are responsible for their plight; AIDS is caused by a moral failure of the individual.[53]

Labeling this perspective as "victim blaming" is to miss its full historical implications. Certainly, it does hold victims of disease socially accountable for their illness, disability, or even, death. But it also underscores implicit cultural values about the nature of behavior, responsibility for disease, and access to care and services. The triumph of a social ideology of individual control marks a powerful denial of our relative lack of control. Because external risks appear so difficult — if not impossible — to modify, the tendency in the last two decades has been to focus on those risks for which individuals do have some modicum of control. Implicit is a subtle psychological defense against the reality that human vulnerabilities and, indeed, mortality ultimately may lie beyond these efforts of individual reform. When running enthusiast and author Jim Fixx died of a heart attack while running in 1984, it was but a reminder of an essential irony that fitness provides no guarantees against sudden, premature, unanticipated death.

At particular historic moments, social forces have eroded the rigid allocation of responsibility for disease. Just as the Great Depression demonstrated that destitution could be the result of powerful forces beyond the control of the individual, so, too, did contemporaneous attitudes about disease reflect an ideology of randomness. The fact that the major voluntary health insurers have their origins in the Depression reveals a social and economic recognition of a shared vulnerability to sickness and its costs. When Blue Shield and Blue Cross were organized in the 1930s they were committed to the idea of community rating: anyone could join at the same cost. The essential philosophy behind this premise was that risks should be equally shared by everyone in the community. In an age of magic bullet treatments and heightened expectations of medicine's ability to cure, political sentiment for the "right" to health care intensified. Calls for universal insurance were one result of this political impulse and, eventually, the enactment of Medicare and Medicaid in 1965.[54]

If individuals, however, voluntarily "take" risks, according to this view, they should be held responsible for the injuries or diseases they incur. As more and more risks came to be identified as a result of the new epidemiology, so, too, did a higher level of personal accountability. As the cultural assumptions about disease causation and responsibility for health changed, so did the discourse on insurance and access to care. By the mid-1970s, discussions of the right to health care had been transformed to the duty to be healthy. In this context, full access to medical care was viewed by some as possibly encouraging unhealthful behaviors. As physician-philosopher Leon Kass notes: "All the proposals for National Health Insurance embrace, without qualification, the no-fault principle. They therefore choose to ignore, or to treat as irrelevant, the importance of personal responsibility for the state of one's health. As a result, they pass up an opportunity to build both positive and negative inducements into the insurance plan, by measures such as refusing benefits for chronic respiratory disease care to persons who continue to smoke."[55] By the 1980s, when new voices supported national health insurance, it was typically to limit rocketing health expenditures—more than 12 percent of the gross national product—than to provide everyone equal access to health care. Equality of access came to be seen as a potential windfall of asserting central control over medical costs.

This chapter focuses on the cultural implications of fundamental demographic and epidemiological change. The way any society comes to understand patterns of health and disease, and their causes, is revelatory of basic cultural norms and expectations. Whether these changes are the result of purposeful human insight and intervention—as they so rarely have been—or of basic changes in the material conditions of life, they do have a dramatic impact on assumptions about human behavior and notions of responsibility and morality. In the United States in the twentieth century, the good life came fundamentally to mean freedom from disease; health became the ultimate goal in a pluralistic, secular culture. And freedom from disease came fundamentally to mean assertion of control over the vagaries of one's body. The ephemeral nature of the effort has yet to discourage this intense capacity for personal reform.

CHAPTER FIVE

Female Science and Medical Reform

A Path Not Taken

Regina Morantz-Sanchez

B EST REMEMBERED as America's first woman doctor, Elizabeth Blackwell was, in fact, much more than just a professional pioneer. She was also self-consciously a reformer and a feminist who saw the practice of medicine as an opportunity for women to bring about fundamental social change. Her ideas about medicine, women, and society bear an uncanny resemblance to contemporary doubts about the alleged objectivity of science, the reduction of medical therapeutics to its biomedical essentials, and the marginalization of women in the medical profession. One cannot help but wonder what the history of American medicine would have looked like if Blackwell's views had predominated.

For example, for more than three decades we have witnessed a mounting critique of science as it is now conceived and practiced. Beginning with the earlier, groundbreaking work of Thomas Kuhn, historians and social scientists have demonstrated how culture has historically influenced the pursuit of scientific knowledge. Meanwhile, feminist scholars have shown that the pursuit of science, particularly since the seventeenth-century scientific revolution, has been a gendered enterprise—namely, an activity not only dominated by men but one dependent on masculine conceptions of reality and masculine theories of knowing. This idea has been a difficult one to assimilate, in part because we, who live in a culture that values the technical, the rational, and the objective above all else, have been taught to consider science the most objective, the most rational, and the most value-free undertaking one can choose.

But Blackwell would have found recent scholarship quite compatible with her own understanding of the effects of the scientific revolution on medicine. The product of a remarkable reform family, including abolitionists Henry and

Samuel Blackwell, their wives, Lucy Stone and Antoinette Brown, and a sister, Emily, who also become a physician, Blackwell developed an approach to her profession that confidently bore the stamp of her family's progressive tendencies. Like many other nineteenth-century physicians who were influenced as much by their moral and religious beliefs as they were by their scientific training, she was a sanitarian in her conviction that there was a social, political, and moral component to sickness and, consequently, to the physician's task. Believing that a benevolent deity had created a universe governed by immutable natural laws, she held that disease was caused primarily by human failings. The modern blights of ugliness, want, and pollution could be banished only through the use of science as a tool for social reform. The individual must be taught the laws of health, while the environment must be stripped of the filth and moral depravity that fostered illness. The good physician addressed not only the health of the body but the health of the body politic.

Blackwell did not invent this approach to medical reform, nor was she alone in advocating social medicine.[1] What is intriguing, however, is that she used feminist language to frame her arguments and that her thinking was, as we shall see, deeply influenced by her conception of gender.

I

Blackwell received her medical training in the 1840s, when the role of the physician was shaped by a traditional system of belief and behavior that saw sickness not as the specific affliction of a particular part of the body but as a condition affecting the entire organism. Illness occurred as the outcome of disequilibrium with the surrounding environment. Body parts were assured to be integrally connected, so that indigestion could be expected to cause emotional distress and depression could produce a sour stomach. Blackwell's training dictated that medical therapy display a visible physiological effect, an effect that could be witnessed and assessed by the patient and his or her family. The science of medicine lay with the doctor's ability to select the proper drug in the proper dose to bring about the proper physiological changes. But the task required art as well: one must be familiar with the patient's unique history and with his or her family's constitutional idiosyncracies, all of which must be assessed. Physicians treated patients in their own homes, emphasizing the sacredness of personal ties with clients and the relevance of family history to clinical judgments.[2]

Though this entire rationalistic framework was labeled scientific, Blackwell and many of her contemporaries understood the word *science* in a different manner than we do today. For example, few physicians in the nineteenth century would have denied the importance of intuitive or subjective factors in successful diagnosis and treatment. "The model of the body and of health and disease," Charles Rosenberg observes, "was all-inclusive, antireductionist, capable of incorporating every aspect of man's life in explaining his physical condition. Just as man's body interacted continuously with his environment, so did his mind with his body, his morals with his health. The realm of causation in medicine was not distinguishable from the realm of meaning in society generally."[3] Blackwell's colleague Professor Henry Hartshorne, who held a professorship of hygiene at the Woman's Medical College of Pennsylvania similar to the one Blackwell originated at the Woman's Medical College of the New York Infirmary, could have been speaking for her when he observed in an 1872 commencement address, "It is not always the most logical, but often the most discerning physician who succeeds best at the bedside. Medicine is indeed a science, but its practice is an art. Those who bring the quick eye, the receptive ear, and delicate touch, intensified, all of them, by a warm sympathetic temperment . . . may use the learning of laborious accumulators, often, better than they themselves could do."[4] At midcentury, when advances in laboratory physiology began to discredit traditional heavy dosing, emphasizing instead the self-limiting quality of most diseases, some practitioners elevated the art of medicine to even greater importance. Skeptical of the utility of intervention, many often sought merely to minimize pain and suffering and to "wait on nature."

The dramatic bacteriological discoveries in the last decades of the nineteenth century led many practitioners to turn to the paradigm of experimental science as a way out of the doldrums of such therapeutic stagnation. Researchers isolated pathogenic bacteria and numerous epidemic diseases — among them cholera, tuberculosis, diphtheria, tetanus, and bubonic plague — and paved the way for the vaccines and antitoxins that gradually became the staples of public health arsenals on both sides of the Atlantic. Yet, despite the high hopes that accompanied these discoveries, Elizabeth Blackwell remained suspicious of their usefulness. Microbe hunters posed three fundamental dangers to medicine as she understood it. First, she saw their conception of the etiology of disease as reductionistic and materialistic. Second, she objected to the moral implications of experiments on live animals, so integral to the new laboratory science. Finally, she feared that a preoccupation with the laboratory could turn the profession away from an emphasis on clinical practice, denigrating the importance of the doctor-patient relationship she held sacred and allowing physicians to neglect their responsibilities in the area of social health and reform.

Blackwell was not the only physician to be suspicious of the new medical reductionism. John Harley Warner and Russell Maulitz have demonstrated how and in what ways laboratory medicine threatened traditional epistemological approaches. While older practitioners continued to emphasize the importance of clinical observation and the inevitability of individual differences in treatment, laboratory enthusiasts argued that the chemical and physiological principles derived from experimentation should be used to inform therapeutics. Patient idiosyncrasies and environmental differences were gradually stripped of their earlier significance, while reductionist and universalistic criteria for treatment took their place. The experimental therapeutist focused less on the patient and more on the physiological process under investigation. The result was a competing definition of what constituted science in medicine and "a thoroughgoing rearrangement of the relationships among therapeutic practice, knowledge, and professional identity."[5]

Such changes prompted Elizabeth Blackwell to devote considerable attention in her later writings to the dangers of this medical materialism. What, she asked, did the pursuit of science really mean? Pure bacteriology never fit her definition of true science, because bacteriologists consistently overlooked the connection between the mind and the body. "We can take a steam-engine or a watch to pieces, examine their parts, repair them, and put them together again," she wrote in 1898. "But a living thing cannot be treated in the same way." To regard human beings as simply "material bodies," without considering the effect of the mental state or the environment on the health of those bodies, was bad science and, consequently, bad medicine. "Sanitary law," she argued, "teaches us that disease is produced by many causes, not solely by a specific microbe."[6]

If disease was multicausational, then the true scientist needed to learn to interpret many kinds of facts. Sanitary science considered evidence broadly; bacteriology did not. "Knowledge," she was fond of pointing out, "is of various kinds: Mental, Physical, Mathematical. These separate departments of knowledge rest equally on bases of fact. Love is as much a fact as bread-and-butter; justice is as potent in its effects as microbes; and from their wider range of action and more permanent duration these mental facts are far more *real* than the physical phenomena."[7]

Blackwell clearly had difficulty absorbing the model of science that gradually insinuated itself into medical practice at the end of the nineteenth century. Its assumptions were that scientists could observe nature objectively, that as they studied natural phenomena they were capable of keeping themselves separate and independent from the object of their research, and finally, that the knowledge one sought by such investigation would be free of specific context and might be reduced to universal laws, generalizations, or rules. Indeed, the careful investigator made an effort to strip away differences in an attempt to find the common elements that bound successive examples together. As John Harley Warner observes regarding the late nineteenth-century laboratory scientist's careful experimentation with physiological processes, "it made relatively little difference whether that process was going on in an Irish immigrant or a laboratory dog."[8] This approach to knowledge had more or less defined research in the physical sciences since the seventeenth-century scientific revolution; the successive discoveries of specific microbes for a range of dreaded diseases eventually demonstrated its utility for medicine as well. According to this model, scientific knowledge was dispassionate, precise, and subject to suitable tests of proof, evidence, and the collective criticisms of the scientific community.

"Science deals with things, not people," Marie Curie observed at the beginning of this century.[9] But if Curie's definition was right for chemistry and physics, Blackwell believed it to be disastrous for medicine. "Science is not," she insisted, "an accumulation of isolated facts, or of facts torn from their natural relations. . . . For science unites and demands the exercise of our various faculties as well as of our senses. . . . Scientific method requires that all the factors which concern the subject of research shall be duly considered." Elsewhere in the same essay she elaborated on what some of those other factors might be: "The facts of affection, companionship, sympathy, justice exercise a powerful influence over the physical organization of all living creatures." For Blackwell, medicine did indeed deserve to be called scientific but not by forcing it within the narrow confines of bacteriology's mechanistic, reductionistic, and deterministic model.[10] The "true" physician had twin obligations to a patient: to cure disease and to relieve suffering. Blackwell understood what some bacteriologists in their enthusiasm were beginning to forget: that it was possible to cure the body through technical intervention while failing to reduce a patient's suffering.[11]

Regrettably, Blackwell was unable to envision the positive contribution such scientific advances would make in the conquest of epidemic disease. Elsewhere, I emphasize the seemingly reactionary aspects of her response and compare it unfavorably to that of more modern-seeming women doctors, like Mary Putnam Jacobi, whose enthusiasm for laboratory science at the end of the century linked her with some of the most revolutionary and dramatic achievements of her profession.[12] But dismissing Blackwell's traditionalism is too easy, because recent critics of medical reductionism have proved her a prescient detractor.

For two decades, psychiatrist George L. Engel has chided his medical colleagues for continuing to adhere to a "model of disease no longer adequate" to its scientific tasks or social responsibilities. "Medicine's crisis," Engel writes, "stems from the logical inference that since 'disease' is defined in terms of somatic parameters, physicians need not be concerned with psychosocial issues which lie outside medicine's responsibility and authority." Like Blackwell, Engel deplores the persistence of a "mind-body dualism" in medical thinking, which assumes "that the whole could be understood, both materially and conceptually by reconstituting the parts." Engel is in a much better position then Blackwell was to acknowledge the extraordinary achievements of scientific medicine in the last century. And so he does. But he also worries that younger physicians are increasingly forced to confront the contradictions between the "excellence of their biomedical background" and the "weakness of their qualifications . . . essential for good patient care."[13]

Indeed, many challengers to the reigning medical dogmatism have ascribed to the ascendancy of biomedical reductionism a health care system that exhibits such undesirable practices as "unnecessary hospitalization, overuse of drugs, excessive surgery, and inappropriate utilization of diagnostic tests." The problem is particularly acute when one reflects that researchers estimate that the medical system (doctors, drugs, and hospitals) affects only about 10 percent of health measured by the usual indexes, with the remainder determined by factors like individual lifestyle, physical environment, and social conditions. In sharp contrast to the attitude of nineteenth-century sanitarians, these are aspects of life over which doctors today have relinquished professional interest. "While reductionism is a powerful tool for understanding, it also creates profound misunderstanding when unwisely applied," H. R. Holman writes. "Reductionism is particularly harmful when it neglects the impact of non-biological circumstances upon biologic processes." Some medical outcomes, Holman concludes, "are inadequate not because appropriate technical interventions are lacking, but because our conceptual thinking is inadequate."[14]

Such contemporary problems have prompted a reevaluation of the content of medical education, an abiding interest of Blackwell's over a century ago. Like modern critics, she too feared that an overemphasis on reductionist science would ill prepare young medical graduates as clinicians. These suspicions led her to mount a vociferous campaign against the handmaiden of experimental bacteriology, vivisection (experiments on live animals) in medical teaching. Finding little support for antivivisectionism among American physicians, her work in this cause was better received in England, where she settled in the 1870s. Here she joined a group of reformers, women physicians, and several prominent male members of her profession to protest the pursuit of knowledge through evil means. She believed that the infliction of pain on living, sentient beings, even to seek good ends, would lead to disastrous professional consequences.[15]

It was not simply the plight of animals that concerned Blackwell, although she shared with many Victorian Darwinists a sentimentalized sense of kinship with them. Her deepest fear was for patients, especially the poor. "The attitude of the student and doctor to the sick poor," she wrote, "is a real test of the true physician." By explicitly encouraging the experimenter to be detached from the object of research, vivisection threatened to harden its practitioners. She believed such teaching would inure students to "that intelligent sympathy with suffering, which is a fundamental quality in the good physician." Soon, she predicted, the sick poor would be regarded simply as "clinical material." She admitted that "these poor victims of our social stupidity are often extremely trying." But the truly attentive physician must never fail to recognize the "real human soul" underneath.[16]

Laboratory research in bacteriology troubled Blackwell for far more than its ethical neutrality, and this brings us to another of her objections to the new science: its neglect of the doctor-patient relationship. While experimental investigators like Louis Pasteur saw themselves as progressives and revolutionaries whose discoveries would rapidly advance the treatment of disease, Blackwell sided with the more traditional members of her profession who still venerated a personal, humane style of clinical interaction oriented toward the individual patient. A new image of the medical scientist emerged at the end of the nineteenth century—calculating, manipulative, and objectively detached from the object of study. To Blackwell, such an individual seemed a wholesale rejection of the interpersonal dimension of medicine on which her entire conception of the physician's task rested.

It is not that she objected to medical research. Indeed, she believed medical advances depended on it. "The present condition of medicine is that of an art, not of a science," she complained in 1898. "There exists uncertainty in diagnosis, uncertainty in the action of remedies, ignorance of individual idiosyncracy, and terrible inability to meet such devastating diseases as cancers, consumption, leprosy, etc."[17] She participated in founding the Leigh Browne Trust in the 1890s, whose purpose was to promote original physiological and biological research "without experimentation on live animals of a nature to cause pain."[18] She strongly supported clinical casework, postmortem and gross pathology, pathological chemistry, microscopic anatomy, and other types of patient-centered investigations. These provided the raw data on which inductions regarding the causes and prevention of disease could be based. "In medicine," she told students at the London School of Medicine for Women in 1889, "anatomy, physiology, and chemistry are the primary studies . . . on which the skill of the future physician so largely depends." In 1889 she wrote, "The conquest of pain and diminution of nervous shock in necessary surgical operations, the disappearance of blood-poisoning, hospital gangrene, and erysipelas . . . are the direct outcome of persevering and skilful clinical observation, of careful work in the laboratory, of humane experiment."[19]

Research, then, was indispensable to the physician's task. But it must be focused on the patient, not on abstract physiological laws and certainly not on the physiology of animals. Like other colleagues similarly contending with changes in medicine, Blackwell emphasized behavior over biomedical knowledge as the basis for professional identity. More important than long hours in the laboratory was a physician's skills in clinical observation and the ability to maintain "character" at the bedside. "It is not a brilliant theorizer that the sick person requires," she reminded students, "but the experience gained by careful observation and sound common-sense, united to the kindly feeling and cheerfulness which make the very sight of the doctor a cordial to the sick."[20]

II

Recent feminist critics of science and medicine have restated in a more sophisticated fashion their difficulties with a model of scientific objectivity that fosters the ethical neutrality that so disturbed Elizabeth Blackwell about vivisection. Blackwell argued that every medical activity contained moral considerations for both doctor and patient, whether it was the simple amputation of a limb, the care of a case of fever, or the birth of a child. One neglected this element at peril to doctors, patients, and society. Today, scientists are capable of performing a whole range of extraordinarily complicated procedures in the area of reproductive technology and genetic engineering, for example, which have the potential for profoundly altering our environment and the structure of our social institutions. Researchers, when they have not been apologists for the outcomes of their work, have generally left it to the public to monitor the applications of their discoveries. We need think only for a moment about the possible applications of creating babies in test tubes to understand what can happen when medical scientists divorce themselves from the social uses of their knowledge.

Moreover, while Elizabeth Blackwell lacked the epistemological sophistication of twentieth-century critics, she cautioned against the "blind acceptance of what is called 'authority' in medicine." She urged her students to remember that "medicine is necessarily an uncertain science" and that "all human judgment is fallible." She encouraged a mild skepticism of "the dicta of so-called medical science . . . skepticism not in relation to truth—that noble object which we hope to approach even more nearly—but skepticism in relation to the imperfect or erroneous statement of what is often presented as truth."[21]

Twentieth-century feminists have taken up this caution and, with the help of contemporary literary textual criticism, have expanded upon it. For example, the late feminist neurophysiologist Ruth Bleier reminds us that scientific dogmatism can no longer exempt the scientific test from the kind of scrutiny being given to other written products of our culture. "Scientists believe [that] the language they use is simply a vehicle for the transmission of information about the objects of their research, another part of the scientist's toolbox, separate from the scientist's subjectivity, values and beliefs. They do not recognize or acknowledge the degree to which their scientific writing itself participates in *producing* the reality they wish to present nor would scientists acknowledge the multiplicity of meanings of their text. . . . As literary criticism has 'debunked the myth of linguistic neutrality' in the literary text, it is time to debunk the myth of the neutrality of the scientific text." The techniques of scientific language (for example, the required passive voice) and the facade of neutrality give the appearance of objectivity and anonymity, which is, in Bleier's opinion, merely an illusion.[22]

Blackwell held that scientific dogmatism could interfere, especially, with the doctor-patient relationship. Indeed, it was usually in defining the parameters of this sacred trust that she drew on her particular brand of feminism. Wedding the nineteenth-century cultural paradigm of woman-as-mother to traditional assumptions about the role of the physician, Blackwell, along with other leaders of the women's medical movement, fashioned a formidable argument for training women in medicine. If the doctor's responsibilities were integrally linked to the family, and if the well-ordered family was to provide a model for a well-ordered society, who better than scientifically trained women could monitor family and social health? "Our medical profession," Elizabeth Blackwell observed, "has not yet fully realized the special and weighty responsibility which rests upon it to watch over the cradle of the race: to see to it that human beings are well born, well nourished, and well educated." Such work "was especially incumbent on women physicians," who, she argued, would be a "connecting link between science and everyday life."[23]

When Blackwell wrote about the role of women in the profession, she often came close to special pleading on the grounds of female biology. Yet she consistently maintained a theory of physician-patient relationships that was not gender bound. She believed that every physician, male or female, must demonstrate what she termed "the spiritual power of maternity." This concept built not only on her formulations of what constituted good science but on her feminist notions of moral responsibility.

Blackwell believed that motherhood, much like the practice of medicine itself, was a "remarkable specialty." But it was not so much "the material aspect" of motherhood that intrigued her. It was a mistake to view motherhood as purely physical, in her view. What made mothering so compelling were the spiritual principles that underlay the ordinary tasks that most mothers daily performed. Indeed, for Blackwell maternity had much in common with Erik Erikson's idea of generativity, a concept that he defined as "the concern in establishing and guiding the next generation." Like Erikson, Blackwell interpreted the "spiritual power" of maternity in the broadest possible terms. Not only physicians, but all mankind, must learn the lessons it had to teach. And what were these lessons? "The subordination of self to the welfare of others; the recognition of the claim which helplessness and ignorance make upon the stronger and more intelligent; the joy of creation and bestowal of life; the pity and sympathy which tend to make every woman the born foe of cruelty and injustice; and hope . . . which foresees the adult in the infant, the future in the present." These, Blackwell insisted, were great "moral tendencies"; they were insights derived from the social practice of mothering, and they could not be measured or reproduced in the laboratory.[24]

In calling for the restoration of the spiritual power of maternity to medical practice, Blackwell was making a case both for the importance of subjective forms of knowing in the clinical encounter and for the broad social responsibility of the profession. Outside of domestic life itself, she wrote in 1891, there is "no line of practical work . . . so eminently suited to . . . noble aspirations as the . . . study and practice of medicine." But that study required the preservation in the physician of the qualities of "tenderness, sympathy, guardianship," qualities that she identified with mothering. These, combined with the more objective methods of scientific inquiry, formed the essence of "true" scientific medicine.[25]

Though perhaps somewhat camouflaged in Victorian prose, Blackwell's ideas regarding spiritual maternity are consistent with the approach to science and to human relationships shared by a group of feminist scholars currently wrestling with many of the problems that she confronted. Although Elizabeth Blackwell operated in a moral and intellectual world where categories of analysis were less flexible and less developed, she clearly identified how and in what ways modern medicine would go wrong. Without necessarily imposing twentieth-century standards upon her, one can find strains in contemporary feminist lore that illuminate Blackwell's thinking and establish her as a significant nineteenth-century intellectual antecedent.

In tracing the evolution of the contemporary ideology of science, recent historical work has described how scientists gradually succeeded in legitimating their approach to knowledge. What has interested feminist scholars most about this process is the way in which the language of science served not only to validate a specific kind of inquiry—controlled laboratory experiments, counting and measuring results, and the publication and replication of protocols—but also to denigrate more subjective and informal modes of knowing. Francis Bacon, for example, one of the founders of modern science, filled his writings with patriarchal metaphors depicting mind as masculine, nature as feminine, and science as a "chaste and lawful marriage" between the two. Nature, devoid of mind, was to be conquered, dominated, subdued, and controlled. Gradually, more subjective forms of knowledge, ones that had played an important role in traditional branches of medieval science like alchemy, persistently became identified with the feminine. Conversely, the systematic social construction of modern scientific activity has been bound up, especially in the last three hundred years, with definitions of masculinity. The result has been a familiar legacy of gendered cultural dichotomies that are still very much with us, dichotomies like male/female, objective/subjective, culture/nature, active/passive, rational/emotional, and public/private.[26]

Contemporary feminist theorists from several disciplines have begun to ferret out the destructive, even dangerous, aspects of such an intellectual legacy, not only for our culture but also for the pursuit of science itself. One approach has been to challenge the socially constructed dichotomies, especially the opposition between reason and intuition. Scholars pursuing this line of thinking have identified and deconstructed the origins of the view that women's close connection with organic life has kept them from the world of reason. Even more relevant to our analysis, however, has been their scrutiny of the relationship between the tasks most women perform in our society and intuitive, more subjective ways of knowing. The growing literature on this subject, though not totally in agreement, can nevertheless give us insight into Elizabeth Blackwell's concept of the "spiritual power of maternity." [27]

In an important article entitled "Maternal Thinking," philosopher Sara Ruddick proposes to revise common assumptions about mothering, which contrast the abstract and formal ideas of men's traditional explorations with the more informal modes of thought involved in child rearing.[28] She takes special issue with an underlying tenet of this dualistic thinking, the idea that because society has relegated mothering to a private, less structured, and more "natural" sphere of existence, a mother's knowledge is not as legitimate or as acceptable as men's knowledge. She proposes instead that a mother "engages in a discipline" just as systematic as the pursuit of medicine or law, that reason is an essential component of a mother's knowledge, and that "maternal thinking" is an activity that deserves to have its forms recognized. Indeed, she goes one step further: she suggests that analyzing and describing this type of thought—a form of knowing shaped by maternal practices—can lead eventually toward the creation of a new public ethic with a greater emphasis on caring.

Ruddick is not alone among scholars who take issue with the presumption that legitimate reasoning occurs only within the realm of "masculine" activities like science and mathematics. Several thinkers show that the devaluation of women's intuition derives from a more general cultural suspicion of all forms of knowledge that are nonscientific in the conventional sense. Moreover, recent developments in a wide range of disciplines demonstrate the inadequacy of using the scientific model as an absolute measure of what counts as knowledge.

Many theorists, male and female, are calling for a redefinition of the notions of subjectivity and objectivity. They argue, for example, that women's activities in the domestic sphere, particularly child rearing, require preparation, foresight, experience, training, and ultimately, the ability to reason. While it is true that most mothers love their children and respond to their needs, that response is not random or automatic. Even maternal feelings are not instinctive but are the product of thought, knowledge, and experience. In the process of living with children, mothers, Ruddick argues, "acquire a conceptual scheme—a vocabulary and logic of connections through which they order and express the facts and values of their practice." This knowledge and experience exhibits a much greater emphasis on the particular and the individual than does that of a mathematician or theoretical physicist. Indeed, because much of mothering entails the fostering of emotional and intellectual growth, a duty that requires responding to a being (object) that is continually growing, changing, and intentionally moving away, a mother's task involves a willingness and ability to gradually alter her relationship with her child. "The idea of 'objective reality' itself," Ruddick observes, paraphrasing philosopher and novelist Iris Murdoch, "undergoes important modification when it is to be understood, not in relation to 'the world described by science,' but in relation to the progressing life of a person." It is this "unity of reflection, judgment and emotion" that Ruddick labels "maternal thinking."

Ruddick's article is one of several recent attempts to critique and ultimately modify the radical separation between subject and object, which has become characteristic of scientific thinking in the last three hundred years. Scholars who share her approach seek nothing less than the incorporation into scientific discourse of a different form of knowing, one that makes effective use of subjective experience. For example, physicist Evelyn Fox Keller argues that objectivity as it has been previously understood is too "static." It begins with the erroneous assumption that one can in fact sever oneself completely from the object of study and, hence, regard one's investigations as unbiased and absolute. She proposes instead an expanded mode of knowing, which she calls "dynamic objectivity." Its aim is to grant "to the world around us its independent integrity" but to do so "in a way that remains cognizant of, and indeed relies on, our connectivity with that world." Keller seeks a special kind of empathy, "a form of knowledge of other persons that draws explicitly on the commonality of feelings and experience in order to enrich one's understanding of another in his or her own right." [29]

Feminist physicians, drawing strength from recent feminist critiques of science, have begun to complain about the ways the traditional scientific method can interfere with humane approaches to patient care. They especially criticize the limitations of a theory of knowing that does not take into account the intersubjectivity between doctor and patient. When the trained physician expects to find objective reality conforming to a specific set of medical theories, when reductionist signs arrived at by mechanical or chemical tests become the sole basis on which practitioners are taught to reach a diagnosis, patients are rendered unimportant as reporters about their own bodies. This is especially true when their subjective experience does not easily conform to the doctor's educated perception of that reality. When pushed, many doctors would likely admit to using another form of knowing in diagnosis when necessary—one that readmits the patient's subjectivity as significant to the healing process. But as one feminist in family practice claims, few would do so publicly. "We have up to the present failed to acknowledge the legitimacy" of this approach, she asserts, "and instead have attempted in our research and our writings to employ what is more accepted as the 'scientific' way of knowing." [30]

Unlike Elizabeth Blackwell, who lived in an intellectual climate dominated by powerful gender dualisms, none of these modern thinkers believe that such a form of knowing is necessarily dependent on women's biology or is even always female. They argue only that women in our society are generally taught maternal thinking, while men are not. In fact, Marxists scholars, who believe thought arises out of social practice, have recently studied forms of work that are most likely to utilize such thinking. Feminist theorists generally tend to attribute women's greater reliance on subjective forms of knowing to our culture's gender-differentiated socialization experiences, especially those of mothering and of being mothered. Others relate such epistemologies to class differences and the experience of subordination.[31] But whatever the origins, the aim is to temporarily abstract such thinking from its social context and measure its potentialities against more traditional methods of scientific inquiry. Feminists offer an alternative paradigm that stresses intuition and the interaction between the knower and the known, arguing that traditional definitions of objectivity actually constrain the future development of science. The goal is an expanded and more effective theory of knowledge, one that can better accommodate not only the pursuit of science but also the development of a more holistic and life-affirming culture.

Blackwell would probably have felt comfortable with modern feminist refinements of maternal thinking. Unlike modern feminists, of course, her notion of the social construction of gender was vague: she was never clear about whether the nurturing she identified with women was socialized or innate. Though I believe she ultimately came down on the side of environmentalism over biology, she did blame bacteriology on the "male intellect" and warned her students against "blind imitation of men" and the "thoughtless acceptance of whatever may be taught by them." She wanted women to learn to question. Yet she also insisted that men could be taught to nurture. "Methods and conclusions formed by one-half the race only," she often pointed out, "must necessarily require revision as the other half of humanity rises into conscious responsibility." Bacteriology constricted conscious responsibility by reducing the patient "to the limits of the senses" and ignoring the fact that each human being was a "soul as well as body." In addition, it obscured "the importance of unsanitary conditions" by playing down "the multiple causes of disease." The effects were already visible: "intellectual ability" was being diverted from "the true path of sanitation by an exaggerated search for bacilli." Worst of all, the profession was losing its social conscience.[32]

At the end of the nineteenth century most medical reformers hitched themselves to the rising star of science. But each age has a unique body of knowledge it considers scientific, and with the ascendancy of bacteriology, scientific reform came to mean something different than it had meant to Blackwell and other sanitarians. The excitement generated by the germ theory led many to exaggerate the importance of specific etiology. Much that had been previously learned about the social and environmental causes of disease was neglected. Elizabeth Blackwell believed health and disease to be social problems calling for political responses. "Pure air, cleanliness, and decent house-room secured to all," she once wrote, "form the true prophylaxis of small-pox."[33] Her fears that reductionist thinking would emphasize the lack of scientific knowledge and promote mere technical solutions has, in many respects, been realized. In place of the extensive social program of the sanitarians, who focused on clean water supplies, filtered sewage systems, adequate food, decent housing, and education, early twentieth-century public health advocates highlighted research and inoculation.[34]

To be sure, the public's perception of the power of science played an important role in medicine's shift toward reductionism. The idea of science became an important component in middle-class discourse. Science was, after all, begetting useful technologies, which gave it prestige in the minds of both intellectuals and the public. Although epidemiologists have questioned whether bacteriology played as important a part in reducing the dangers from infectious disease as improvements in living standards, what was most important was the public perception that medical science was responsible. Bacteriological interventions appeared to prolong life.

Medical professionalization reinforced these changes, because reductionist definitions of science helped raise the status of physicians. Unlike holistic approaches to care that often breached the wall between doctor and patient and depended on a body of shared, accessible information, the new scientific model mystified medical knowledge and furnished objectively agreed upon criteria for setting standards in behavior, diagnosis, and treatment. Increasingly, the profession advocated social relations—between itself and the public, between itself and other "healers"—that kept it insulated from alternative views of practice that might illuminate and improve health care.

In clinical settings, bacteriological thinking had an important effect on the way medicine was organized and delivered. The practitioner was gradually taught to devalue subjective forms of knowing in favor of technological expertise. It is not that the doctor denied the importance of caring in treating a sick patient. Indeed, nurses were expected to perform such services. The reductionist argued only that hand-holding was not necessary to produce results. More and more, subjectivity would be viewed not as an alternative mode for gathering essential data but as an error or even a contaminant in exercising judgment.

By the end of the nineteenth century, most women physicians welcomed bacteriology and vivisection, anticipating their positive effects on medical practice. But they continued well into the twentieth century to offer a morally distinct alternative to the prevailing vision of scientific medicine that was gradually emerging. Their overrepresentation in social medicine during the Progressive Era suggests that Blackwellian concerns with holistic approaches to medical care continued to be part of their professional vision well into the era of experimental science, technology, and specialization. Even as they welcomed new technologies, many continued to voice Blackwell's concerns regarding the duties of the profession. Eliza Mosher, for example, an emerita president of the American Medical Women's Association, lamented in 1925 the passing of the "human" in medicine and regretted the loss of "the sympathetic relation which formerly existed between doctors and their patients." In Blackwellian overtones, she warned her colleagues, male and female, to beware of "narrowing and concentrating their vision upon the purely physical to the exclusion of the psyche and human." [35] Other spokeswomen continued to believe that women physicians in particular must exert their influence in favor of preserving holistic approaches to treatment.

But women physicians have until very recently been marginalized in twentieth-century medicine. Perhaps partly as a consequence, the issues Blackwell raised—in medicine, in science, in feminism—remain unsolved. Certainly, the exclusion of women from a significant role in twentieth-century medical professionalization has meant the loss of an epistemology often, though not solely, derived from the primarily female activities of nurturing, generating, and caring. What has been reinforced instead is a narrow and rigid cognitive stance that has confined subjective forms of knowing increasingly to the realm of art and literature and to the private world of the family. As subjectivity has been feminized and devalued, its effective use has been undermined even for the women who have successfully entered the profession of medicine. The art of medical care has been relegated to nurses, and women doctors who pay too close attention to it are often viewed as unprofessional. Instead, the supposedly gender-blind concept of scientific objectivity, modeled on the rigid separation of subject from object, knower from known, has informed and dominated approaches to illness and patient care. Patients are distanced from practitioners, and doctors are distanced from the social responsibilities of their profession.

In addition, the absence of women has facilitated the organization of medical education and practice in a fashion best suited almost entirely to the male life cycle. The demanding routine of call schedules and residency requirements has penalized or completely eliminated doctors wishing to participate in a full and satisfying personal life. Such developments have certainly reinforced the exclusion of women, but they have also effectively barred the male physician who sincerely believes that a person with family ties, human emotions, and diverse intellectual interests makes for a more effective healer.

Though we cannot measure the specific costs to medicine of the loss of

Blackwell's voice, we can be reasonably certain that her approach would have addressed these issues and attempted to solve problems that our modern, technological culture has overlooked. Somehow, Blackwell understood what a century of dealing with the shortcomings of reductionist medicine has only recently led thoughtful contemporary thinkers to apprehend—that being a humane and sympathetic practitioner is essential to being a competent scientist. Without such qualities, the physician's task of gathering all relevant information cannot be accomplished.

PART III

Selling Science

THE STORY of science in the twentieth century is partly the story of its location and distribution. Increasingly, science was dispersed, both in the sense of being practiced in a wide variety of government and private institutions and in the sense of being widely used as a source of authority, language, and metaphor. Yet some kinds of scientific practice were also concentrated by their dependence on a limited number of funding agencies and by the necessity of marshaling complex resources in a central place. One result of the latter was a sharper boundary between science, as specialized and arcane knowledge, and the world of everyday life and common sense. For much of the century, that separation reinforced the image of science as above it all — politically neutral, disinterested, and authoritative. The chapters by John Stilgoe and by Roland Marchand and Michael Smith demonstrate the simultaneous dispersion and concentration of science as well as the difficulties, consequences, and social costs in distributing it to the public.

Stilgoe discusses a "time-tested tradition of farmer as experimenting scientist," which the U.S. Department of Agriculture largely ignored after 1890. Small farmers in particular had their own standards of proof and efficiency. In the end, they found government science less congenial than their own knowledge or what they received from corporations that were sensitive to their needs. This is a cautionary reminder that "local knowledge" can be a form of science and a powerful counter to the authority of official science. As in other essays in this volume, the tale is one of the need for openness, debate, and respect for differences that pass the tests of experience and public scrutiny.

Marchand and Smith examine corporate efforts to publicize and promote admiration for their scientific research, arguing that—in John C. Burnham's words—from the 1930s onward there was a "displacement of emphasis from knowledge to product." These corporate campaigns, most visible in World's Fair exhibits and reaching a pinnacle in EPCOT, were naturally designed to sell consumer goods, but in the 1930s they also had an anti–New Deal agenda and denied "conventional notions of social or political reform." In doing so, they denigrated history, perhaps indirectly supporting the point made by Hollinger, Ross, and Kloppenberg that historical thinking can be a powerful tool of reform. Although attempting to tie consumerism to a narrative of scientific progress, these corporate displays increasingly found it hard to maintain that faith in the face of intractable social problems and of criticism of the sort discussed in this book. Progress may have remained the most important product of consumer industries, and science may have been its parent, but by the end of the twentieth century it had become a much harder sell.

CHAPTER SIX

Plugging Past Reform

Small-Scale Farming Innovation and Big-Scale Farming Research

John R. Stilgoe

I N 1920 THE United States Department of Agriculture announced its twentieth-century vision of reformed farming. "Certain well-founded principles are involved in the planning of an industrial establishment, which apply as well to the laying out of a farm plant, and there are others which are peculiar to the farm problem," argued the authors of Farmer's Bulletin 1132, *Planning the Farmstead*. "It is the purpose of this bulletin to explain the principles affecting the farm layout by applying them to given problems."[1] Unlike so many other farmer's bulletins, *Planning the Farmstead* is generalist in orientation. Rather than provide specifics about barn siting, drinking water location, and other essential elements of the nucleus of farm buildings composing the farmstead, two "rural engineers" struggle to reshape farmers' thinking. Principles, not common sense, tradition, or hard-won experience, must govern thinking, and planning must reshape the farm "plant." A new vocabulary carries the gospel of scientific management along with an older message of agricultural science.

Efficiency long worried the editors of agricultural periodicals and other savants determined to rescue American farmers from backwardness. Indeed, the agricultural press of the newly independent nation championed change for greater efficiency even as it celebrated independence, and the championing continues unabated today. But late in the nineteenth century, subtle forces reshaped the thinking of many American farmers, forces that prepared some farmers for the vocabulary of Farmer's Bulletin 1132, even for its opening remark that "farming is a business, just as is the manufacture of automobiles or any other industrial activity," forces that caused other farmers to discard the bulletin along with other government-issued "trash" and to look to other sources.

Manufacturers of farm machinery, fertilizer, tools, custom feed, even building materials instructed farmers in the uses of their products and in the industrial philosophy that produced them. From advertising and engineering offices in Chicago, Pittsburgh, and other cities flowed the so-called principles of science and technology that farmers, in time, often accepted as their own. Supposing themselves independent, free-thinking, the backbone of the Republic, farmers acquired industrialized science and technological thinking along with the industrial products and the free hats emblazoned with manufacturer names and slogans. Sometimes voluntarily, sometimes unknowingly, sometimes in desperation, they sifted a variety of offerings and often chose private over public information. When they chose private industry science over government-funded science, they often did so not only because manufacturers addressed those farmers ignored by government agencies but also because private industry science honored the time-tested tradition of farmer as experimenting scientist.

Of course, by 1920 the Department of Agriculture had convinced itself that it orchestrated the advancement of agricultural science and was well embarked on a fierce effort to convince congressmen, journalists, and above all, farmers that its efforts epitomized disinterested effort. In his USDA Yearbook, 1920 essay, "Science Seeks the Farmer," yearbook editor L. C. Everard wrote glowingly of federal home demonstration and county agents. "They are salesmen of science, and the wares they have to offer are the combined knowledge and experience of the army of scientists and practical agriculturists of the State colleges and departments."[2] In his preface to the volume, a brief essay entitled "More Complete Knowledge," he insisted that increased agricultural productivity would result only from "applying more intensively on the farm and in the forest the scientific principles and methods that come forth from laboratory, sample plot, and experimental farm." But not just any laboratory, plot, or experimental farm would do, as the last two lines of the preface make clear. "In the long run, however, more complete knowledge of production and marketing, emanating from scientific and unbiased agencies, will go a long way toward solving the problems of producer and consumer alike. The key with which to open the door to better conditions may take any one of a number of forms, but it must be cast chiefly from the metal of Agricultural Science."[3] Home demonstration and county agents sold high-quality science, which was synonymous with government agricultural science, science that deserved capitalized letters — almost official science.

Department of Agriculture official science documents flooded rural America after 1890. In 1920 the department had in print more than five hundred farmer's bulletins offering "practical suggestions and information," along with department bulletins providing "new information obtained by the scientific staff through research and investigation." Other documents included circulars offering emergency information, the *Weekly News Letter,* the weekly and monthly crop marketing reports, the *Journal of Agricultural Research,* the monthly *Experimental Station Record,* and a range of publications about weather. Amid the lesser documents regularly appeared the massive annual yearbook, which by 1920 was 888 pages long and jammed with everything from immediately useful articles like "Phosphorous in Fertilizer" and "Cows that Make the Income Climb" to general essays like "Flowering and Fruiting of Plants as Controlled by the Length of Day" and "Uses of the Soil Survey" to hundreds of pages of agricultural statistics.[4] The gargantuan publishing program derived largely from educational intent, although some of it — the *Journal of Agricultural Research* begun in 1913, for example — existed to connect the work of USDA scientists with other researchers worldwide.[5] Yet, clearly the program competed with others, those programs that Everard considered biased.

Today the public effort receives sustained scholarly attention. The year 1862 endures as the year of the Homestead Act, the Morrill Land-Grant College Act, and the act establishing the Federal Department of Agriculture, acts of immense importance in the history of the Republic. Subsequent public effort acts are less well known but almost as important. The 1887 Hatch Agricultural Experiment Station Act funded largely practical (at first) research at research centers controlled by land grant universities in each state and territory, and the Smith-Lever Act of 1914 provided for a nationwide network of federally funded county agents to "extend" advice and information. Subsequent federal legislation strengthened public agricultural research, and until the National Science Foundation and the National Institutes of Health deflected 1960s attention and resources, the great agricultural effort remained by far the flagship of government-sponsored scientific research.[6]

Farmers responded to this effort with less than enthusiasm. As scholars have only recently demonstrated, the farmers who consistently favored research and education were the prosperous, large-scale ones.[7] Having the capital, or the courage to borrow it, such farmers applied new technology immediately and enjoyed—sometimes—immediate increases in productivity or total output or both. Called progressive farmers in USDA literature, these late nineteenth- and early twentieth-century men appear to have had better formal education than many of their neighbors and may have had a clearer perception of farming as an enterprise rewarding adaptation, not traditional effort.[8] Moreover, after World War I they understood themselves as more closely linked to each other by region, crop specialization, farm size, openness to change, or other characteristics than to the general population of farmers.[9] Accepting innovation and willing to abandon previously successful strategies the moment they began to be unprofitable, progressive farmers championed public agricultural research, especially the close-to-home research of federally funded state experiment stations.[10] In so doing they earned the grudging admiration and open dislike of the "pluggers."[11]

Pluggers were farmers "not very strong on book larnin' stuff," who through continuous hard work, attention to detail, and above all, common sense prospered—or at least endured. Pluggers understood farming as a way of life in which "the farm is the greatest and most important of all factories" *and* the family home.[12] Undereducated, undercapitalized, such farmers avoided risk whenever possible and saw agricultural research and progressive farmers sometimes as models but often as destroyers of economic equilibrium.[13] Not until the 1920s did the pluggers see hard evidence that increased efficiency and, especially, output often lowered commodity prices. After 1920, as progressives and pluggers alike endured a seemingly endless era of depressed prices, crop infestations, and other agricultural trouble, pluggers often came to resent book-learned "big" farmers and their associates.

Pluggers regularly perplexed government-funded experts. Not only did they often disregard research published in bulletins or demonstrated by agents, they often persisted in raising crops and livestock no longer particularly profitable in their regions. They held onto techniques and machines long after government-funded experts had devised replacements. Gradually, the great government-backed research and education effort began to disregard them, then ignore them. Sharecroppers, tenant farmers, market gardeners, part-time farmers, and other agriculturists too poor, too isolated, or too stubborn— essentially, too unredeemable—drifted into a sort of limbo late in the nineteenth century, a limbo unvisited except by the occasional private industry salesman or expert who knew, or guessed, something about the half-hidden prosperity of many pluggers and something about how pluggers viewed science, something about how the nonprogressive farmer, alone in his field, alone in his barn, alone with his problem of the moment, thought.

"The arena of country life, labor, and struggle, where the farmer and his family achieve their primary habits of thought and action, is the farmstead," argued rural sociologist Charles Josiah Galpin in the opening sentence of his 1918 *Rural Life*.[14] In and around the farmstead, children learn not only by precept, not only by doing, but also by being continuously immersed in activity, in farming. Unlike the early twentieth-century city child, who lived away from his father's business, the farm child grew up in the midst of an all-encompassing environment, one typically stretching beyond the family farm into a rural neighborhood of similar farmsteads and farms. And while the mixed-agriculture farmer must be sometimes soilsman, sometimes dairyman, sometimes engineer, and while "the necessity of shifting mental gears from one role to another on short notice creates an intellectual situation which calls for careful consideration," in the end the farmstead acted as a nearly unchanging reinforcement of traditional practice and traditional thinking. "The handed-down methods of farm practice, upon the handed-down farmstead, with its handed-down horizon line and landscape outlook, are no inconsiderable forces in shaping rural psychology and rural organization."

Galpin discerned the importance of the farm and farmstead as a "theater of operations," in the military sense, and this term explains, perhaps, the intensity of Farmer's Bulletin 1132, which advocated revolutionizing the theater: scramble the farm and then physical, and perhaps mental, gear shifting will proceed more smoothly; perhaps science and technology will transplant quicker. The term also explains the explicit bias of Farmer's Bulletin 1132 toward farmers of the Midwest, the progressive farmers so beloved by the USDA, the farmers who had the resources — financial and intellectual — to reshape their farmsteads, to change their horizon lines. Pluggers adapted to handed-down farmsteads and "made do," resisting official science and technology.

Somehow, for all his apparent stubbornness, the plugger kept alive the early nineteenth-century notion of the farmer as inventor, the farmer as scientist, until by 1924 one of the largest tractor and implement manufacturers, Deere and Company, suggested that book-learned farmers looked to the plugger for good advice in hard times. According to Deere, the plugger thought for himself, tested for himself, and knew quality when he saw it, whatever the opinion of others. In the first decades of the twentieth century, no element of the farmstead announced more clearly the triumph of plugging than the bars closing openings in fences and walls.

Gates

In May 1853, an East Barnard, Vermont, farmer reported the results of a home-bred scientific experiment, the sort of effort that shaped much twentieth-century enterprise. J. Davis alerted *New England Farmer* readers in a letter so laconic and pointed that the journal editor praised it as a "model communication" and one "worth a 'mint of money,'" for its evidence "is strong." [15]

In 1838, the letter recounts, Davis "took a stick 14 feet long and cut it in the middle, setting the butt of one up and the other down, 12 feet apart." After five years, "the one with the butt down rotted off." At the date of his writing, "the other stands sound yet." More than chance or casual happenstance informs his letter. Davis used the same piece of wood, not two similar-shaped pieces of the same species—say two roughly equivalent pieces from two different chestnut trees—or even of the same tree. Moreover, he placed the two posts cut from the "stick" only a dozen feet apart, apparently to ensure similar, if not perfectly identical, soil conditions. And finally, he recorded the time elapsed before rotting. Davis meant his post setting as an experiment. "I had heard it stated that the top end of the stick should be stuck in the ground." He tested tradition and new knowledge.

The letter prompted lengthy editorial comment, summarizing the most recent findings in the national debate about the rotting of fence posts. On the one hand, the editor remarks that some farmers simply "noticed that some of the posts remained nearly sound, while others were rotted off at the bottom," and went "looking for the cause." But on the other hand, farmers had begun deliberately experimenting. One "set a series of white oak, and for the purposes of testing the theory, set every other one top part down." Another set two chestnut gateposts one butt down, one top down, and observed the results. The editor guessed that sap vessels might have one-way valves and that setting posts butt up made the rise of ground moisture difficult. "But we are not certain about this," he concluded. "The subject is important, and worthy of the attention of some of our scientific correspondents. All persons making fence will do well to remember these facts." His final words balance perfectly on-farm science with off-farm science. Experiment—rigorous, controlled, and repeated—demonstrated "facts" to be accepted by all farmers, but only off-farm science could explain them.

Davis wrote in the midst of the great fence debate, a debate as multifaceted as any in agriculture but focused on efficiency. Discussion on the sturdiest fence, the cheapest fence, the longest-lasting fence posts, the strength of gates led to assertions and analyses of efficiency. Arguments concerning gates and bars depended on the most careful experiments: advocates of gates—board affairs swinging on hinges affixed to posts—scorned advocates of bars—two or three rails set horizontally between two posts.

As early as 1833, champions of gates were attacking the owners of bars as backward. "A farmer must be rather an awkward man who cannot make a common farm gate, and a dull mathematician who cannot calculate the advantages of using them instead of bars, in all places where the business of the farm requires frequent passing and repassing," asserted the *Genessee Farmer* in an article reprinted in the *New England Farmer*.[16] Mathematics shaped the argument: a man and team take five minutes longer to pass through bars than through a gate, and assuming "he only passes and repasses once each day for one half of the year (as bars are generally left down one half of the year), this would amount to one hour each week or three and a quarter days in each year." Given the value of a man and team as $1.50 a day and the cost of a home-made gate as $4.65, the gate would pay for itself in about one year. Gates, not bars, were not only "economical" but "add much to the general appearance of a well-conducted farming establishment." By 1842 champions of gates had convinced most editors of agricultural periodicals, and their claims became increasingly stark—and vague: "The time saved in passing through gates, instead of pulling down bars and fences, will amount to many days in the course of the year."[17] In the agricultural press from 1850 on, gates, not bars, mark the well-conducted farm.[18]

Gates depend on gateposts, however, and gateposts presented a variety of difficulties, which prompted letters from Davis and other farmers determined to learn the causes of rot, sway, and other vexations—and to eliminate them. Farmers exchanged their experiences throughout the nineteenth century, producing by 1900 an intricate and wonderfully rich literature reporting on-farm hypothesis testing. Should posts be set with or without bark removed? In 1835 one farmer, who had "not found suitable leisure till this day, which is by far the most tempestuous known in this region for the last thirty years," set down his "own experience" in testing bark-stripped posts smeared with tar.[19] A year later another reported from New York on the outcome of "some experiments made" to forstall gatepost rotting. He argued that gateposts must be made of well-seasoned wood, of course, but what really mattered was the filling of the holes. If the hole is filled with stones no larger than five inches in diameter, and "a thin mortar, or grout" is poured around the stones, the gatepost will be not only firmly anchored but also far less likely to rot.[20]

Within a year, the *New England Farmer* published experiences demonstrating that "a post planted when green will last longer than one previously seasoned."[21] An 1839 *Farmer's Cabinet* writer argued that cedar makes the best posts, but what worked best against rot was simply setting big posts, especially white cedar.[22] The outpouring of letters prompted further experiments, often reported years after the letters that sparked them. In 1842 the *New England Farmer* printed a letter by a farmer who had improved an 1836 idea by rounding the top of the mortared stones to shed water from his eight-by-ten-inch gatepost. After two years the gatepost stood perfectly solid in a block of concrete "two foot square."[23] While such reports of experiments duplicated or modified appeared intermittently throughout the century, innovations entered the literature too. By 1850, one Connecticut farmer had convinced himself that cast-iron posts lasted best, and within nine years a New York farmer provided readers of the *Country Gentleman* with a detailed comparison of the qualities of cedar and cast-iron posts, "letting a few facts speak for themselves."[24] But cast-iron posts signaled a change in the literature — the arrival of "patent" posts, fencing, and gates.

Well into the middle of the nineteenth century, letter-writing farmers reported building solid gates wide enough for a team and wagon to pass through for about fifty cents cash outlay for the hardware.[25] Hinges, even the most crude staple-and-ring type, meant a visit to a blacksmith and a cash transaction, and farmers tried to invent gates using the simplest hinges but remaining free from sagging. "I look upon a farm without gates as behind the age, and would almost as soon think of building a house without a door, as attempt to manage a farm without a gate at every point where frequent passage to and from enclosures is necessary," asserted one Ohio farmer in an 1859 *Country Gentleman* letter. W. C. Pinkham insisted that gate making defeated many farmers "who are intelligent on all other subjects" and who needlessly hire "mechanics" to build gates — which then turn out to sag or sway.[26] His sketch and text carry his argument that, with blacksmith-made hinges, good quality wood (including black locust for the posts), and precision of measurement, "any farmer with skill enough to hold a plow can make such a gate, at expense of not more than one dollar."

Pinkham took pride in his plan and urged farmers to test it. "Farmers, try one of these gates; it will cost but a trifle, and at the end of one, five, or ten years, report through the *Country Gentleman* the result of your trial." Pinkham's challenge hit home, for experimenting with lines of force, particularly the force of gravity, intrigued farmers anxious to use gates rather than bars but equally anxious to spend little money on heavy hinges and iron reinforcing rods. "I have seven field gates now, and I need as many more, and I should very much like some *cheap* yet *durable* plan," wrote one Missouri farmer in 1859.[27]

From the high ground of completed experiment, he criticized Pinkham's gate offered only a few months earlier in the *Country Gentleman*. "It has never yet ceased to swag," he wrote of the sagging, swaying gate built according to plan. He offered a design "dispensing with iron hinges" and much more adequately braced. The following week the journal published another letter offering different improvements to prevent sag.[28] An Illinois farmer demonstrated in text and sketch his scheme, which depended in part on iron hinges but allowing thinner uprights. Within several weeks appeared a letter from the originator of the flurry of letters, Pinkham himself, stressing that his gates had lasted a year "without sagging a particle I can discover" and that several built on an earlier, less adequate plan had lasted five years without sagging.[29] Pinkham emphasized that he had done his experiments well and that his diagonal braces made sense in their position. "The best test of its merits, however, is a trial, which I recommend with much confidence," implying that the letter writers had botched their trials.

Farmers broadened the discussion throughout the Civil War, largely as a result of the general victory of gate champions over adherents of bars but also out of growing concern with the cost of mechanic-made gates and factory-built products. Correspondents began reporting twelve-year-old nonsagging gates, and editors began suggesting that simple, tested gates could be built in two hours and trusted for years.[30] But after the war, letters changed; correspondents began asking for information on or comparisons of patented gates, even the most simple.[31] Interest in manufactured fencing, posts, and gates had grown since the 1850s, especially with the advent of iron posts and wire fence and the appearance of complicated, patented gates guaranteed against sagging.[32] Yet until past the turn of the century, correspondents and essayists returned to the old questions involving bars and farmer-made gates. "When I see a farm well supplied with gates, I know that man is progressive and making money, for he values time, and wisely seeks to save it," argued John M. Stahl in an 1884 *American Agriculturist* piece aimed at helping farmers make nonsagging gates fourteen feet wide.[33] "In these days of self-binders," Stahl decided, no farmer could afford gates too narrow for the new machinery, but making wide gates required experiments in counterweights, hinges, and other components different from those of earlier decades. Farmers who chose to build their own gates rather than buy manufactured ones had to rethink old information about strengths of wood, stresses, and fasteners, let alone setting gateposts; and farmers who bought gates had to consider not only first costs but maintenance expenses as well.

Acute observers of the turn-of-the-century rural landscape discerned a contrary approach to the gate design issue, however, one that illuminates the complexity of farmers' responses to innovation. Beginning in the 1880s, writers in national magazines began labeling as backward those farming regions displaying little in the way of gates or manufactured fencing, especially woven-wire fencing. They scorned the snake fences of the southern uplands and the stone walls of New England, even as they admitted that wire fencing had not justified the "great hopes" of reformers, and they noticed with mixed amazement, nostalgia, and scorn the continuing presence of bars.[34] A pattern, vague and at times scarcely discernible, orders their comments and suggests that, innovation notwithstanding, many farmers remained pluggers, at ease with bars, perhaps even happy with them.

Contemporary observers of rural scenery, life, and customs rarely paused to consider the day-to-day workings of the farms they described, and almost never did they interview farmers. They never learned that bars endured decade after decade for straightforward reasons that no amount of letter writing and "trials" could change. The rotting of gateposts progressed fastest in wet areas, and farmers retained bars at pigsty openings and other muddy points simply because gateposts and gates deteriorated quickly, something that escaped the notice of many observers who never noted that wet areas boasted more bars than gates.[35] Heavy snow defeated gates everywhere, however, especially in hill pastures open to drifting snow. Gates swing, and require arcs clear of snow. Farmers sledding home firewood knew the wonderful advantage of bars dropped quickly into the snow: dropping the bars meant a delay of a minute against the ten minutes or longer needed to shovel out a gateway. Homemade, cheap to replace, useful in snow and high wind, never sagging, bars endured for the sensible reasons dear to pluggers. But something more diminished the impassioned arguments against bars and the scores of letters recounting experimental gate design, something that passed almost unremarked in its time. Some farmers, especially dairymen and farmers keeping several milk cows as a sideline, stepped back from the arguments and experimented with eliminating or reducing the number of bars and gates altogether.

Essentially, they rearranged their farms to make their fields connect not to each other but to "common lanes," fenced pathways known in regional English as driftways or barways. Sometime in the 1870s, this concept fired the enthusiasm of efficiency-minded farmers aware not only of the roundabout routes needed to drive cattle from barn to pasture but of the stupidity of using whole fields as rights-of-way rather than as cropland. While colonial New Englanders had reserved "rangeways" along the edges of fields they sold to others, the barway derived from a wholesale restructuring of the mid-nineteenth-century farm.[36] In the name of efficiency—and to accommodate ever-wider horse-drawn machines—farmers relocated stone walls, not only making fields bigger but ordering them along skinny, walled ways called lanes or, since farmers closed them to cattle grazing in some contiguous pasture, barways.

In 1873, the *Illustrated Annual Register of Rural Affairs* notes that "a moment's reflection must show that a well-planned sub-division of a farm lies at the very foundation of convenience, system, and economy" and glimpsed the wonderful by-product of replacing "the needless furlong" with a barway.[37] In orienting his fields along one barway, the farmer dramatically decreased his dependence on gates and bars, making old-fashioned, homemade bars far more efficient. After decades of dispute, the old bars had proven efficient in a wholly unlooked-for way. Farmers who had invested time and money in sophisticated gate building learned the frustrating permanency of gates, while bar owners enjoyed the flexibility of moving bars easily to new locations. Pluggers found that, once placed in slightly different, less frequently used spots, the old bars worked more efficiently than poorly sited gates. Pluggers laughed last.

The great bars-versus-gates controversy of the nineteenth century suggests the range of the debates that surfaced in agricultural periodicals, state agriculture reports, and agricultural society bulletins. It also reveals farmers' willingness to experiment and the diversity and sophistication of that experimentation, what one speaker at an 1859 agricultural society meeting called "observation and experiment."[38] James J. H. Gregory told his Essex County, Massachusetts, listeners that "knowledge in agriculture appears as but a phantom, flitting in the dusky twilight and eluding our most earnest efforts to secure her in our grasp and fix her habitation," while demonstrating through example after example that it would be farmers themselves, publishing their observations and experiments in the several agricultural "papers," who would in the end make the snatch.

The controversy demonstrates, too, the richness of experimentation outside of government-sponsored effort, indeed before there was much government-sponsored effort. This experimentation was "sponsored" by farmers themselves and, after the 1870s, also by manufacturers of gates. And it also demonstrates the final triumph of pluggers over progressives. In the 1870s and 1880s, farmers finally figured out that the core issue was not the relative efficiency of bars and gates but the proper placement of each type of barrier. One farmer might need six sets of bars and one gate; another might need twelve gates and one set of bars. Until each farmer knew his needs—and especially his need for rearranging his fields—choosing "progressive" gates over "old-fashioned" bars might do nothing to increase convenience, system, and economy. By 1920, the presence of bars on a farm indicated, to government-sponsored research-ers and their progressive allies, sloth or ignorance or both. But to others, they indicated a thorough, determined, wait-and-see farmer, the one machinery manufacturers courted, the plugger.

Wheel Hoes

Government-backed researchers tended to ignore a great late nineteenth- and early twentieth-century transformation in American agriculture—the rise of the small-scale, often part-time farmer. By 1900, the big-agriculture angle of vision had nearly blinded Department of Agriculture officials and research-ers to the existence—let alone the needs—of thousands of prospering but hard-to-categorize families dependent on agriculture for part of their income. Sharecroppers and other tenant farmers received sustained scrutiny largely because USDA researchers knew that such men worked for large agricultural operations, especially on former southern plantations or on one-crop mid-western farms whose owners supported government-sponsored research and made decisions for the tenants. But the small-scale farmer presented all sorts of difficulties.

As early as the Civil War, eastern agricultural publications noticed the "one-horse farmer" and from time to time published letters and articles intended, as one letter writer put it, as "vindication" of the "class from the prejudice that is apt to attach to any 'one-horse concern.' " [39] The anonymous letter writer un-hesitatingly accepted the mid-nineteenth-century eastern embrace of intensive agriculture ("a little farm well tilled proves more profitable than a great one under hurried, superficial cultivation") and argued that men with other in-come sources—clergymen, physicians, tradesmen, teachers—are the farmers who innovate most because they know their operations intimately. Even with his other job, the one-horse farmer can "avoid the risks incident to farming on a more extended scale" and often finds that "his crops are put into the ground and have even grown high enough for their first hoeing, before farmers in gen-eral have finished the needful preparation for planting." About the size of the one-horse farm, the letter writer said little, hinting at a place of six or ten acres in area, assuming, along with other contributors to periodicals, general knowl-edge of the "class" and its operations and noting the confusion of terms like *small farmer* and *market gardener.* Yet a tension characterizes his letter, a ten-sion originating directly in his understanding of machinery.

The writer assumed that small-scale and part-time farmers embraced ma-chinery as thoroughly as they embraced intensive cultivation and that they had enough work to justify machinery, especially the horse-drawn mower. But did they have enough land? Or did they "hire out" with their machine, doing custom work for large-scale farmers desperate to get their hay mown, for instance? The writer suggested that they did, that one-horse farmers, "if benevolently inclined, lend them a helping hand." Was it their success or their smallness or their part-time effort that accounted for the prejudice against them that the writer noted? Was it their outside income that allowed them to buy the small machinery, the one-horse machinery, that boosted their output and made their custom work attractive to large-scale farmers? Were they somehow out of the mainstream of agriculture, at least of agriculture as seen by official government sources?

Again, the answer lies in the nineteenth century, in this case in the form of patents. New England and New York farmer-inventors addressed the failures of large-scale agriculture in their region by creating machines peculiarly adapted to the intensive cultivation championed by reformers and other visionaries, a cultivation that was beginning to show real profits. Their machines demonstrate a first-hand knowledge of farming coupled with a strong drift away from agriculture and toward manufacturing. One publicist of their efforts, the New York–based journal, *The Plough, the Loom, and the Anvil: Farmer and Mechanic,* offered a venue for the growing union of farmer-inventors and craftsmen-manufacturers. Inventors advertised machines of little use in Ohio and other midwestern states, machines like the "hand cultivator" patented by New Hampshireman Jonathan A. Robinson.[40] The cultivator, essentially a wheelbarrowlike, two-wheeled device fitted with knives that straddled a row of plants and cut weeds as its operator pushed it along, was "designed for garden or field cultivation." According to the editor, the inventor "has hoed small carrots with this machine, the points of the cutters being 1¼ inches apart, and he walked right along, hoeing them perfectly."

Robinson's hand cultivator deserves attention chiefly because Robinson pushed it. No horse pulled it along. One of many inventions created for small-scale agriculturists, it stands far in the shadows cast by McCormick's reaper and other large, horse-drawn machines that midwestern and western farmers, agricultural editors, most manufacturers—and almost all government-sponsored expertise—boosted with such enthusiasm throughout the second half of the nineteenth century, the machines that now epitomize agriculture-focused invention and that shape inquiry into farmers' adaptation to change. But refined versions of diminutive machines like Robinson's made possible the successes of mechanic-minded, one-horse—and no-horse—farmers and announced the triumph of research undertaken almost in spite of federal policy.

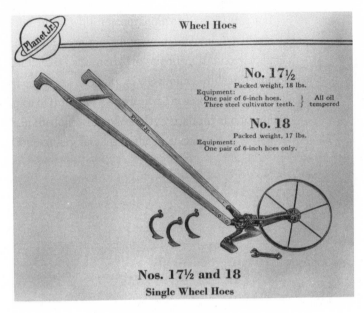

FIGURE 6.1. S. L. Allen & Co. Wheel Hoe Diagram
(from collection of the author)

By 1890 the Philadelphia manufacturer of one-horse farming tools known as the Planet Jr. Line had proven beyond doubt the nearly insatiable market for small-scale farming implements. (See figs. 6.1 and 6.2.) S. L. Allen and Company prospered by discounting USDA precepts, and it continually enlarged its niche in American agricultural manufacturing. Its success demonstrated the inherent genius of Jonathan Robinson and other midcentury inventors of devices useful to pluggers.[41]

Essentially, Allen and Company prospered by selling lightweight, multipurpose machines that farmers pushed. Gradually, the firm expanded into a line of tools pulled by a single horse, and in the first decades of the new century it manufactured not only implements designed to be pulled by small gasoline-powered tractors but also its own tiny, two-wheeled, engine-driven garden tractors guided by walking farmers. From its beginnings in the late 1870s, the firm focused its efforts on recreational gardeners as well as one-horse farmers, flourishing as suburbanization gathered force, of course, but succeeding, too, by serving a large group of farmers all but ignored by government-backed researchers and the large-implement manufacturers of Chicago, Moline, and

FIGURE 6.2.
S. L. Allen & Co. Wheel
Hoe Advertisement (from
collection of the author)

other midwestern cities. But a powerful devotion to experimentation shaped company policy, too, and may well explain company successes.

While the firm consistently emphasized that its head "has always been and is today a practical farmer himself, and therefore in position to know what farmers want," continuous "careful experiment in the field" lay behind the continuous refinements in the tools. The firm conducted trials on its own test plots, but it brought customers into its efforts by manufacturing and selling more than one version of a tool and asking customers buying the hopefully improved version to report back to the company. In 1892, for example, the firm offered only the new version of the "combined drill" (a device that planted seeds automatically), having sent out the year before some of the new drills, "with special requests in each box to report to us after trial." Precisely how the policy worked cannot be determined, but the rapid design changes and frequent references to customer trials suggest that the firm intended its catalogues as educational literature as well as sales brochures. Simply put, Allen and Company had to assume the difficult burden of convincing one-horse farmers, who were scorned by government-backed experts as hopelessly hidebound, that a new tool would prove profitable. Moreover, the convincing proved complex, for not only did it involve generalities, it also involved all the specifics of every machine and accessory. And the firm determined that it must entertain suggestions from its customers. As the decades passed, its catalogues evolved from fairly straightforward, detailed product descriptions to booklets describing an alternative, largely metropolitan agriculture of small holdings and few implements.

Throughout its history the firm emphasized its push wheel hoes and automatic seed drills, devices equipped with all manner of gadgets and practically impossible to describe briefly. Wheel hoes came with one wheel or two: single-wheel machines typically ran between rows of plants and sliced off or tore up weeds, while two-wheel machines straddled one row of plants and sliced off weeds on either side. Drills deposited seed in perfect intervals, in one operation opening a furrow, depositing seeds ranging in size from tiny (celery or onion) to large (bean) at intervals set by the operator, and covering and tamping the dropped seeds, all the while marking the next parallel row. The firm claimed that with wheel hoes "one can plant four times his usual acreage of hoed crops from drilled seed, without fear of being caught in their cultivation." Fourfold efficiency increases depended on the seed drill as well as the wheel hoe, of course, and Allen and Company manufactured combination tools like the "Combined Drill, Wheel Hoe, Cultivator, Rake, and Plow," an 1892 near-top-of-the-line tool. "Every purchaser of this machine will find it an excellent seed sower; a first-class double-wheel hoe while plants are small; a first-class single-wheel hoe; an excellent furrower; an admirable wheel cultivator; a capital garden rake; a rapid and efficient wheel garden plow, and it is without an equal in a variety of tools, easy adjustment, lightness, strength and beauty, and as a practical everyday time and labor saver." Allen tools had very specific applications, indeed, and in the end, that proved important to the man—or woman—with five acres of celery to plant and cultivate.

It proved of importance to children, too. Allen and Company understood the role of children in the small fields of the one-horse farm, and it created smaller-sized machines for young pushers. The Planet Jr. single-wheel hoe "is light and well suited to the use of boys or girls," the firm claimed, emphasizing that the height of the wheel could be changed "to suit the depth of work and height of the operator combined" and that attachments were easily inter-changed, "a great advantage to beginners or when the tool is placed in unprac-ticed hands." But the model had other uses. "In some very tough soils beaten down by rain, it is even profitable sometimes for a *man* to use but a single hoe and go rapidly along." In the end, the boy ought to use the exceptionally lightweight Fire-Fly wheel hoe, "a good tool for the boys and a pleasant one, though a thorough, strong, all-day tool for a hard-working laborer," for after all, boys lacked the energy reserves of grown men.[42]

The firm understood that most of its customers owned a horse, usually a general-purpose animal not particularly powerful. But the horse plowed and harrowed the small fields that, thereafter, were worked by pushed tools, and if the small-scale farmer owned too much arable land for a wheel hoe or two, he might buy one of the firm's "one-horse cultivating tools." Diminutive by West-ern standards, the "tool can be used in the most delightful manner for hoeing a crop closely, saving an immense amount of work in all crops which are usually hoed by hand. It is particularly useful to the market gardener and trucker, and to broom-corn growers, and, in fact, to all who grow crops where hand work must be done." Accessories and different models enabled growers to cultivate beneath leafed-out plants, in vineyards, and along rows of sweet potatoes. One very special tool, the celery hiller (first marketed in 1888), enabled growers to throw earth onto celery and so to blanch it.[43] But whatever its innovations, until late in the 1890s the firm intended its tools to be pulled by one horse.

After the financial panic of the early 1890s, Allen and Company entered the larger-farm market with a two-horse, pivot-wheel cultivator, but in its 1897 catalogue it explained that its recent efforts had focused chiefly on developing labor-saving tools for financially straitened small-scale farmers and market gardeners. For the first time, the firm began emphasizing large market gardens owned and operated wholly by women—perhaps by women whose husbands were forced to work full-time away from the land.[44] Within a few years the firm had become established in the medium-scale-farm market, but it continued its emphasis on one-horse farmers until massive technological change forced substantial reorientation.

By 1920 many Allen and Company customers no longer owned horses. In-stead, they owned automobiles for commuting to work and, sometimes, for delivering their produce. And very quickly the customers learned the difficulty of being without horses in springtime, when plowing had to be done when weather and soil conditions permitted. This moment in American agricultural history, a moment reaching across a ten-year period between 1918 and 1928, passed unnoticed by government-backed experts but not by Allen and Company. By 1925 it was shifting its emphasis to recreational gardeners unfamiliar with wheel hoe technology, explaining in a free booklet, *A Good Garden is Half of a Good Living,* how to manage a very large home garden with little effort — and a combination wheel hoe and seeder.[45] Of course, West Coast competition had begun to destroy the established operations of eastern metropolitan market gardeners, and the firm moved swiftly to extol the healthful exercise resulting from pushing its machines — anything to expand its recreational gardening market. Throughout the 1920s its catalogues edged toward being how-to books aimed not only at experienced market gardeners and other small-scale farmers but at newcomers — both hobbyist and profit-oriented venturers. Yet not until 1930 did the firm perfect its ten-year researched response to horseless one-horse farmers.

Its garden tractor eliminated all need for horse-powered implements, at least on land already in cultivation, and sometimes eliminated the need for pushed machines, too. The garden tractor boasted a noisy gasoline engine riding between two large, spoked, metal wheels and pulling the same attachments the firm sold with its pushed machines, although made of steel rather than cast iron (the garden tractor snapped iron attachments when it crashed into hidden rocks). Proud as it was of the machine, the company often advertised it negatively, explaining what it was not. "The Planet Jr. is not an all-purpose tractor," it argued in 1930. "It will not pull a large plow through heavy sod. It will not do the type of cultivation expected of a large tractor." What it did was replace the light-duty horse. "It is primarily a cultivator of narrow-row vegetable crops, and for that purpose is second to none." Most growers, the firm admitted bluntly, used it "to do the work of wheel hoes," and a few determined large-scale growers used several of them in concert to work fields of a hundred acres of spinach and other crops, apparently because the men "guiding" them tired much less quickly than if they pushed wheel hoes, even the Fire-Fly.[46]

Around 1930 the firm realized the diminishing resources of so many of its customers and the fierce competition of the tiny gasoline-powered cultivating tractor on which farmers rode, especially the International Harvester Farmall and its diminutive successor, the Farmall Cub. A one-horse farmer might never buy a garden tractor, or might buy a riding tractor, or might stick with an old horse until its death put him out of business. Many market gardeners simply sold their acreage to real estate speculators, and many hobbyist gardeners simply did not need an engine-driven tractor of any sort.[47] In the depths of the Depression the firm abandoned its horse-drawn and engine-driven products and struggled to keep its decades-old pushed products for sale to those pluggers who understood them to be what they had always been: low-cost, productivity-increasing machines.

However important the tens of thousands of wheel hoes sold by Allen and Company and its competitors, they are remarkably absent from government-sponsored research and educational literature. Indeed, their absence is almost extraordinary. Not only did experiment stations and other branches of the increasingly massive federal- and state-supported system publish nothing devoted specifically to wheel hoes, they rarely mentioned wheel hoes in reports on crops for which wheel hoes were ideally suited. No government-backed expert attempted to improve the wheel hoe and wheel hoe accessory design, or to evaluate models made by competing manufacturers, or perhaps most important, to instruct farmers and pleasure gardeners in their use. The vast late nineteenth- and early twentieth-century crop-growing literature barely mentions wheel hoes.

Once in a great while, government-supported researchers did cite the advantages of wheel hoes. In its 1893 *Strawberries,* the Agricultural Experiment Station of the Rhode Island College of Agriculture and Mechanic Arts aimed its advice at three groups of readers: large-scale farmers, pleasure gardeners, and small-scale farmers.[48] For the large-scale farmers it offered advice about using a horse cultivator to kill weeds, and for pleasure gardeners it explained the use of a hoe, but for the market gardener it offered detailed information, including a line drawing, on the wondrous advantages of a wheel hoe. "The best hand cultivator that we have used for this purpose is the 'Success,' manufactured by Messrs. Kirkwood, Miller & Co., of Peoria, Ill. The wheel of this cultivator is thirty inches in diameter, with its bearings nearly directly over the center of the draft." Everything about the machine was efficient. "It is easily operated and its work is very satisfactory. With it one man can often kill as many weeds and loosen the soil about the plants as nicely as three or more men with hoes." Was it more satisfactory than Allen and Company machines? Apparently so, for in singling out the Success, the Rhode Island Experiment Station implied that the large-wheel machine triumphed over the small-wheel variety manufactured much closer to home.

In the decade in which Americans embraced the bicycle, the big-wheel/little-wheel controversy about wheel hoes may have seemed of value only to one-horse farmers. Yet clearly Allen and Company believed its small-wheel design to be far superior to anything developed by its competitors. The Philadelphia firm marketed its wheel hoes with fourteen- or fifteen-inch wheels because small wheels let operators get very near seedling plants without injuring them.[49] But large-wheeled machines, those with wheels two feet or more in diameter, pushed far more easily and, by the 1920s, appear to have equaled the popularity of small-wheeled ones, although the many attachments of Allen wheel hoes enabled the company to dominate the market.[50]

What explains the rarity of statements like that made by the Rhode Island Experiment Station? Government-supported researchers willingly evaluated privately manufactured products, from fertilizer to tractors, publishing nitrogen, potash, and phosphorous analyses and horsepower ratings, so clearly the researchers might not have feared offending wheel hoe manufacturers, especially those located out of state.[51] The only plausible answer lies in the increasingly big-farm bias of research and the progressive farmers' support of increased experiment station budgets.

Berry growers, however, used wheel hoes. Allen and Company and other manufacturers understood berry growing to be a labor-intensive, usually small-to-medium-scale, season-intensive enterprise ideally suited to wheel hoe use—exactly the viewpoint of the Rhode Island Experiment Station. Yet only rarely did experiment station experts address the needs of small-scale, possibly part-time, berry growers. Liberty Hyde Bailey, perhaps the best known of turn-of-the-century agricultural writers, argued from his Cornell University Agricultural Experiment Station that berry growers should use hoes or horse-drawn cultivators.[52] In *Blackberries*, an 1895 bulletin, he told commercial growers to space their rows eight feet apart, for this "allows of easy cultivation." But cultivated by what? "Two horses and a spring-tooth cultivator are the most efficient means which I have yet found of keeping a blackberry plantation in condition," Bailey concluded enthusiastically. The two-horse method is the whole focus of the bulletin, and Bailey directed the attention of his reader to "the picture in our plantation, on the title-page," an illustration making exceptionally clear that his intended audience consisted chiefly of large-scale farmers.

A national figure in agricultural experimentation and education, Bailey now and then mentioned wheel hoes. His *Principles of Vegetable-Gardening*, a very popular textbook first published in 1901 and frequently revised, lambasted the hoe as "a clumsy and inefficient tool" and praised the "important" wheel hoe.[53] "It saves immensely of hand labor and usually leaves the soil in better condition than does hand-work." But immediately after this, his prose becomes vague. "There are a number of patterns, large and small. Choose a large wheel with a broad tire, that it may ride over lumps and travel on soft ground." About the benefits of the broad-tired, small-wheeled machine he said nothing. Indeed, his final sentence suggests a very real confusion. "Soil must be in good condition to be worked with wheel hoes; therefore, they should be introduced for their educational effect." Whatever the meaning of his last words of the subject—educate about tilth or wheel hoe efficiency or what?—his treatment of the machines in a college-level textbook aimed chiefly at aspiring farmers seems at first inexplicable. Why nearly ignore a machine so useful to small-scale growers?

An answer lies toward the end of *Principles of Vegetable Gardening*, in an outline listing tools for market gardeners. Bailey listed wheel hoes in two places, under "tools to prepare the land for planting" and "tools for subsequent use." But it is on a second list of equipment "for a market garden large enough to be worked by horses or mechanical power" that he lavished attention, suggesting that a would-be market gardener ought to have two horses at least. His book ignores wheel hoes because it ignores the small-acreage market gardener, an agriculturist even the USDA admitted made a comfortable family living from as little as three acres.[54]

By the close of the first decade of the new century, a handful of pleasure gardeners largely outside the circle of government-backed researchers had determined that vegetable gardens ought to be redesigned to make the most of wheel hoe efficiency and that the wheel hoe ought to be the center of a larger "system." Systematizing vegetable gardening in order to increase efficiency struck no one as new, but the notion of the wheel hoe—and to a lesser extent, the scuffle hoe—as the generators of a system did. While Allen and Company had long claimed that large-scale truck farmers frequently employed ten or more men on large fields, each pushing a wheel hoe, not until 1911 did one innovator study the efficiency of the wheel hoe in the small, family-sized garden, a space thirty by sixty feet, "slightly over one twenty-fifth of an acre." E. L. D. Seymour argued in "Economy and the Vegetable Garden" precisely the opposite of Bailey's two-horse thesis and demonstrated his argument through the most precise bookkeeping imaginable, accounting for such minuscule sums as the cost of row labels.[55]

Seymour claimed that his eighteen-hundred-square-foot garden ought to provide all the vegetables needed by a family of four and ought to consume very little time if engineered for the wheel hoe. Using the industrialist-engineer language of the times, he stressed the necessity of eliminating "waste" and emphasized that "economy is simply a synonym for the prevention of waste." Seymour loathed wasted space and wasted produce, but above all he loathed wasted time and energy. Only the wheel hoe, around which the space of the garden is ordered, conserved time and energy. The "adoption of the system of planting in rows instead of in small, isolated beds" allowed the gardener "to weed and cultivate rapidly, down one row, up another, with no breaking of backs and wearing away of knees, merely by the propulsion of a wheel hoe or the rapid manipulation of a scuffle. These modern gardening tools—the second factor in economizing time and effort [after the arrangement of long rows]—should have a prominent place in every garden." He told readers to lavish care on the tools, to keep them in perfect condition, to learn to use them well. "Maintain system in every phase of the work," he insisted.

His photographs and sketches make clear that his "combination wheel hoe and seeder" was indeed "indispensable in the vegetable garden" and "saves time and backache." Moreover, the illustrations depict what are almost certainly Allen and Company machines—small-diameter, one- and two-wheel machines for working close to seedlings. Seymour admitted the necessity for some hand weeding, but he suggested that the use of system and wheel hoes would nearly eliminate kneeling. And his precise calculations of cost—including the property tax on the garden area and the nine-dollar cost for the wheel hoe seeder—suggest that the no-horse gardener could profit from the systematized vegetable garden. Moreover, his article implies that an even larger area would hold out a promise of increasing family income—not just a saving on the food budget.

By the 1940s, Depression and war convinced even government-backed researchers that wheel hoes made sense in any small-farm financial equation, but by then the two-wheel, engine-driven garden tractor and its successor, the rototiller, had reduced many of the claims of wheel hoe supporters.[56] Yet between 1910 and 1940, thousands of American families more than proved Seymour right, although their efforts at market gardening innovation passed essentially unnoticed by USDA and other "official agricultural science" authorities until the Division of Subsistence Homesteads of the Department of the Interior funded several studies in the mid-1930s. In study after study, state agricultural experiment stations examined the extent and implications of part-time farming, learning, as W. P. Walker and S. H. DeVault noted in *Part-Time and Small-Scale Farming in Maryland,* that "small farms are generally overlooked in most agricultural programs. We have neglected to consider small farms in our public policy. Irrespective of the trend toward large-scale, mechanized farming, small farms, like other small business enterprises, are here to stay in sufficient numbers to be a factor in our land policies."[57]

Surprise spices all of the studies. Experiment station workers made discovery after discovery, and while rarely as outspoken as Walker and DeVault, often expressed their wonder that so much small-scale and part-time farming prospered beyond the notice of decades-old agricultural research activities. Iowa researchers found part-time farmers enjoying many more "urban" conveniences, especially radio sets and toilets, than their full-time counterparts; Maryland investigators marveled that "the amount of food that can be secured from a small area of land is surprisingly great." Indiana researchers discovered that many part-time farmers bought a horse in spring, worked it until harvest, then sold it, saving the cost of keeping it over the winter, and that many, many others (72% of the survey sample) used no horses at all.[58] The researchers had burst into an agricultural realm, and often quite a prosperous one at that, about which they knew very little, a realm not only of one-horse farmers but of half-horse farmers, no-horse farmers, and wheel hoe farmers, innovators beyond the notice of brand name agricultural science and technology.

Rocks

How did the national, government-backed agricultural science establishment lose sight of the small-scale, part-time, wheel-hoe-pushing farmer, the plugger who used bars instead of gates? And how did that farmer, market gardener, truck grower—whatever his precise designation—evaluate ideas, technologies, and tradition?

One possible answer to the first question slumbers now in the masses of USDA literature produced just before the turn of the century, in documents whose tone suggests a gradual, subtle shift in USDA research. In a *Yearbook, 1897* essay, "Every Modern Farm an Experiment Station," Ervin E. Ewell told farmers that "science is to the modern civilization what the heart and circulating blood are to an animal" and that "as each heart beat sends a wave of blood that extends to the minutest capillary vessels, so each new discovery goes out from the little group of investigators as a mighty pulsation that sends an enlivening thrill throughout the entire civilized world—a thrill that is felt the more strongly by the individual the nearer he is in education and training to the authors of the discovery."[59] In Ewell's metaphor, farmers might be numb toes. Despite the best efforts of the government-backed scientific establishment, he argued, many, perhaps most, farmers do not receive easily or quickly its discoveries, nor are they quick to apply them. If they distrust reported results, he emphasized, they need only do the recommended experiments, exactly as scientists specify, and prove the worth of professional scientific inquiry.

No thinking could be further removed from that which James Gregory propounded years earlier before his audience of farmer-experimenters.[60] Ewell spoke of experiments as mere demonstrations, as educational clinchers, not as actual inquiries into the unknown, and he said nothing about any farmer making an original discovery. This attitude, in time, made the plugger, and especially the small-scale plugger, invisible to the ever-larger agricultural research enterprise supported by public funds. Ewell understood progressive farmers as those who willingly embraced the results of USDA-supported research. In time, government-funded agricultural research focused its efforts almost exclusively on those farmers it called progressive, typically large-acreage owners eager for the latest government bulletin.

This research, however, even lost sight of some large-acreage farmers, especially those working "substandard" ground. Despite the existence of state experiment stations supposedly focused on the peculiarities of local farming, over the decades following 1880 the public-funded research effort shifted its attention from crops and stock of lessening importance to farming specialties "suitable" to particular states and regions. In New York and Vermont, for example, experiment stations focused their efforts on increasing milk production, in Maine, potato farming, and in New Hampshire, poultry. If the articles in farm bulletins are any indication, farmers choosing crops or livestock specializations different from the mainstream ones of their state or region received increasingly less attention from their state stations as well as from national-level research, and their county extension agents often provided them with advice based on out-of-state information.[61]

Rocks, especially the big rocks eastern farmers called niggerheads, stymied the efforts of many agricultural experts and tried the patience of thousands of eastern farmers.[62] As the westward movement brought thousands of farmers into the rockless Eden of the Mississippi basin, and as implement manufacturers based in midcontinent labored to devise larger and larger horse-drawn machines for the big, rockless fields of the West, eastern farmers and small manufacturers struggled to work around rocks.

Now and then, they succeeded. Walter A. Wood produced a spectacularly successful two-horse riding mower wonderfully adapted to the rocky fields of New York and New England. (See fig. 6.3.) In 1866 his machine triumphed at the Paris Exposition, winning, in the words of Charles L. Flint, secretary of the Massachusetts Board of Agriculture, "not only the grand gold medal for perfection of work and another for superiority of mechanical construction, but also the decoration of the Imperial Cross of the Legion of Honor," all in all "a triumph in which every American must feel a just and natural pride." At county fairs throughout New England, Wood's jointed bar mower won ribbon upon ribbon, proving itself before throngs of farmers eager to abandon scythes but determined to spend no money on second-rate equipment. By 1868, Wood had sold a hundred thousand of the machines, largely because

FIGURE 6.3. Walter A. Wood's Jointed-Bar Mower (from collection of the author)

the lightweight, rugged mower did one thing splendidly—it worked over and around rocks.[63] Of course, all its engineered sophistication meant little west of New York, where mowers encountered few rocks and where giant manufacturing firms competed for the bulk of the implement market. A mowing machine intended for rocky fields sold only to farmers already facing failure—and receiving little government help to continue their hay- and grain-raising efforts.

After the Civil War, eastern farmers abandoned the rocky, hilly fields on which few horse-drawn machines worked well, if at all, and struggled to make their bottomland "pay." Many who remained in farming turned from field crops to sheep raising, then dairying, following the dictates of market returns and the suggestions of government-backed experts. Others tried desperately to make western and midwestern technology work but almost always failed. Rural New England barns still shelter machines sometimes a century old but scarcely used. The mowing machines—called horse killers—were too heavy for sloping fields, and the potato diggers scooped up bushels of fist-sized stones along with potatoes. Such machines often bankrupted "progressive" farmers determined to mechanize.[64] Other farmers drifted into part-time farming and out of the range of government-funded research. They bought wheel hoes; others, a few others, bought dynamite. "Boulders and rock ledges offer the same troubles as stumps in fields," argued the anonymous authors of *Explosives in Agriculture,* a 1931 handbook published "to define proper methods and procedures to be employed in using explosives in agricultural work."[65] Boulders and stumps, in the opinion of the private industry experts, "take up valuable room and reduce the yield per acre; prohibit the use of improved machinery and increase the cost of cultivation; break and destroy tillage equipment; harbor vermin, plant diseases, and weeds; are unsightly and seriously detract from the cash value of a farm."

Since the turn of the century, manufacturers of explosives, and especially the E. I. DuPont de Nemours Company, instructed farmers in the many agricultural uses of dynamite. After 1910, DuPont made an intensive effort to sell dynamite to farmers and quickly understood that top-quality instructions must serve as the core of its advertising campaigns. It published a range of guidebooks, from general to specific, explaining how farmers—and their wives (for after all, the product was safe enough for women to use)—might use dynamite to make tree holes and ditches, eliminate rocks and stumps, and kill woodchucks. But other efforts supplemented the publication effort. Moving pictures, demonstration fields, traveling salesmen, free calendars combined to further farmers' purchase of dynamite. In this way, DuPont created the "agricultural blaster," a farmer who did custom blasting for other farmers. He worked for himself, not DuPont, although he purchased DuPont products and followed, the firm fervently hoped, DuPont instructions appearing in its magazine, the *Agricultural Blaster*. The firm believed that blasters would teach farmers the ease and efficiency of DuPont products and that farmers would start blasting for themselves. Blasters eliminated stumps and boulders, to be sure, but more important, they gave farmers a chance to make their own experiments.[66]

Experts at the Institute of Makers of Explosives understood something that Ewell and many other government-backed experts missed—that every farmer faced peculiarities in his operation, however slight, that required experimentation. Moreover, both DuPont and the institute understood that its customers often owned poor quality land, that they were pluggers, that they were not prosperous, bottomland farmers. The literature they put out encouraged, within limits, individual experiment. "Every stump is a problem in itself," argued the explosives experts. "The age and kind of stump, nature of the soil, ground conditions, time of blasting, character of the root system and equipment available all must be taken into consideration when determining the best and most economical method of removing stumps."[67] In the booklets published by the institute and in the directions published by its best-known member, the DuPont Company, one message was clear: whatever the discoveries of the experts, every farmer must find his own way, for every boulder, every stump posed unique challenges.

At first glance, government-funded experts argued in the same way. "Every job of plowing presents a problem of its own, and there can be no one best method for all cases," argued H. R. Tolley in a 1919 farmer's bulletin entitled *Laying Out Fields for Tractor Plowing.*[68] But his well-illustrated, detailed booklet wholly ignores farmers confronting the rock-filled fields so precisely described and depicted in booklets published by dynamite manufacturers. Tolley's fields were the great, square, rockless ones of the Midwest, not those of the Northeast, the upland South, or the mountain West. Nowhere in his bulletin did he tell a farmer what to do when faced with a rock, and he devoted only three pages to discussing irregularly shaped fields. Not surprisingly, within a few years the chief implement manufacturers had begun publishing massive tomes explaining the intricacies of plowing with tractors.

Soil Culture and Modern Farm Methods, a Deere and Company publication written by W. E. Taylor, chief of the firm's soil culture department, described in its 651 pages state-of-the-art tractor farming—and a mass of other information too.[69] Far surpassing in detail the firm's fantastically popular *Operation, Care, and Repair of Farm Machinery,* a clothbound book of 256 pages that passed through seventeen editions by the mid-1930s, *Soil Culture* argued that "the farmer must be the judge." Given the land and the equipment, the farmer would experiment. *Soil Culture* opened with a poem, "The Plugger," and a statement that the firm had learned much from its customers

Aloneness

On January 22, 1887, the front page of the *New-England Homestead* featured a front-page, illustrated article entitled "Gates Are Cheaper Than Bars."[70] (See fig. 6.4.) Written by an Alleghany County, New York, man, apparently a farmer, the article suggests using horseshoes to hinge a new sort of barrier, half-bars, half-gate, and certainly inexpensive. Many decades after its beginning, the old argument had plenty of energy. Farmers continued to experiment, continued to plug along. Such plugging along entranced some early twentieth-century observers like Charles Josiah Galpin, who was by 1924 head of the USDA Division of Farm Population and Rural Life and who devoted part of his *Rural Social Problems* to analyzing "why farmers think as they do."[71] Galpin decided that the "solitary business" of the farmer contributes, perhaps shapes, his peculiar habits of mind. "The farmer thinks and makes up his own mind. This is because his mind does not clash with those of others. He has to make up his own mind by himself so much that he is not accustomed to the social process of quiet conference and modification." Alone, or with his sons—who learned

GATES ARE CHEAPER THAN BARS.

Perhaps the simplest and cheapest farm gate is one much used in this section, and made as shown in the illustrations. As will be seen, this gate is made the same as a fence

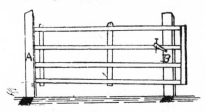

panel, and it is opened and shut by lifting the free end, sliding it and then swinging it around. At A a horse shoe is nailed to the post to hang the free end of the gate on when shut. At B is a pin made of two inch oak

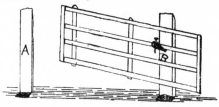

FIGURE 6.4.

An 1887 *New-England Farmer* depiction of a farmer-invented bar-gate hybrid (from collection of the author)

plank set in the post at an angle of 45°, or, in other words, cornerwise. The outer end of the pin is about three inches wide with an iron pin put upright through it, as shown in the cut.—[E. M. Wilson, Alleghany county, N.Y.

from him and, therefore, were unlikely to offer new suggestions—the American farmer thought his solitary way across his farmstead and farm. Galpin concluded that "modern science" often had little discernible impact on farmers, not only because its results failed to reach many of them but also because many of them simply disregarded the results. Along with others in government-backed agricultural research, Galpin championed intensified communication through publications, extension agent visits, and changes in the schooling of farm children. He did not suggest an inquiry into the reasons so many farmers tossed aside USDA bulletins or a reorientation of its research effort so that the department would learn from farmers, particularly from pluggers, who succeeded in unlikely places with unlikely produce or livestock.

The solitary thinker tinkering with a seed drill or changing the diet of hens or noticing a remarkably early-bearing tomato plant and saving its seeds or mixing his own fertilizers or placing dynamite sideways had the ear of private industry, however. Implement manufacturers, feed and seed firms, fertilizer makers, and others listened, not only to gain an edge over competitors but also to offset their lack of research-and-development departments. Little remains of their listening, however, for private industry records are scarce. What endures and shapes almost all historical inquiry are the records of official, government-backed agricultural research aimed at progressive farmers, those who reformed correctly, obediently, those who rearranged their farms and farmsteads, those who abandoned solitary thinking.

And what of the plugger, with his old-fashioned farm and farmstead layout, his bars, his lonely decision making, his almost studied refusal to progress further than his father? He is indeed the unreformed, his farm noticeable still in the landscape, his thinking nowhere recorded. But now his once-scorned discoveries, techniques, and results shape the awesome reform movement sweeping across American agriculture, a movement designated by such awkward terms as *organic farming* and *sustainable farming*. The movement already boasts its own national magazines, which are filled not with the findings of government-supported agricultural science but with the tests of farmers prospering by contradicting USDA advice.[72] These farmers are doing things in traditional and novel ways, are plugging—and are writing about their efforts in a vernacular way remarkably like that of the nineteenth-century gatepost letter writers.

At the close of the twentieth century, the pluggers have proven the value of nurturing both traditional and innovative farming practices, of choosing from a larger menu of information than that provided by government offices. They continue to puzzle both government bureaus and private industry. They are as articulate as the farmer letter writers of the nineteenth century, as free to innovate as any one-horse farmer eager to prosper in changing times, and as well educated as professional foresters and other university-trained experts. The pluggers of American agriculture still, routinely and regularly, test science in the field.

CHAPTER SEVEN

Corporate Science on Display

Roland Marchand and Michael L. Smith

EARLY IN Mark Twain's *A Connecticut Yankee in King Arthur's Court,* Hank Morgan confronts a problem in display strategy. As a foreman for a late nineteenth-century arms factory, he possesses scientific and technical knowledge undreamt of in Camelot. He could tap the educational potential of his various feats (extinguishing the sun, destroying Merlin's tower with lightning, restoring the waters of a sacred fountain) by explaining the principles and devices that made them possible (his knowledge of a solar eclipse, lightning rods, hydraulic pumps). Or he could pass them off as magic, with elaborate incantations and fireworks to dramatize his wizardry. Hank's unhesitating choice is to present science and technology as magic, thereby ensuring his rise to power (and ultimately, his defeat). Replacing Merlin as the court wizard, he becomes a master of special effects: to outward appearances, all that has changed is the more businesslike title Hank chooses for himself: the "wizard" becomes "the Boss."

By the time of Twain's death in 1910, a new breed of science-based corporations was emerging, dwarfing the factories of Hank Morgan's 1880s both in scale and in complexity.[1] By the 1920s and 1930s, electrical, chemical, automotive, and communications companies faced their own version of the Connecticut Yankee's dilemma: should they attempt to educate the public regarding the research and production processes behind their products or present themselves as Aladdin's lamps, magically applying science and technology to fulfill the needs and wishes of consumers?

At the end of the 1920s, as scientific research laboratories bulked larger and larger within corporate strategies for development, merchandising, and public relations, Arthur Ruel Thompson of the Publicity Bureau of Bell Telephone Laboratories confessed the difficulties in making corporate science visible to the public. Research was "rarely picturesque," he lamented. Laboratory scenes provided "unpromising materials for illustration." As a remedy for this barrier to dramatic publicity for corporate science, Thompson proposed the pictorial creation of an " 'atmosphere' setting." Visual imagery for the public would suggest "the idea of research" but would have "no concrete relation to the actual work described." Through a "series of subterfuges . . . not deceptions," he proposed, "the scientist and his laboratory will become wholly fanciful, symbolic entities." [2]

Thompson posed the dilemma of the visual representation of science with respect to a single media—the poster. The swelling parade of engineers, designers, and public relations experts who devised the exhibits of corporate science for fairs and expositions in the following decades worked in a far less constricted medium, one of three-dimensional displays with opportunities for audience interaction. Still, for a variety of reasons beyond the problems of pictorial representation, they moved inexorably in a direction similar to that proposed by Thompson—toward strategies of presentation that were increasingly fanciful and "atmospheric." If, as John Burnham persuasively argues, the largely corporation-generated popularization of science since the 1930s has led to "a displacement of emphasis from knowledge to product," it has also brought a transformation in strategies for the display of science and technology from process to aura.[3] Increasingly, exhibits at fairs, expositions, and theme parks have abandoned demonstrations of process and the elucidation of underlying scientific principles in favor of entertaining and undemanding journeys into corporate "worlds." What corporate exhibitors have not abandoned is the powerful imagery of pioneering and "new frontiers." Although the contributions of science to "progress" have increasingly depended on corporate research teams that work in intellectual realms ever-more inaccessible to the average citizen, the verbal and visual imagery of pioneering has continued to dominate corporate display. This individualistically charged aura of frontiersmanship—which the exposition visitor is often invited to share through some trivial form of participation—ironically continues to suggest the sufficiency of science and technology as the answer to any perceived societal need for reform.

Many forces shaped the progression from process to aura in the corporate display of science and technology. Science itself rapidly became more complex, less amenable to commonsense explanations. The scientific and technological dimensions of the burgeoning chemical and electrical industries were largely invisible to people in their everyday lives. And when corporate experts scrutinized public behavior at fairs and expositions they found short attention spans, easy distractibility, and expectations of undemanding entertainment. At the same time, major corporations were coming to appreciate how a reputation for scientific research could serve them in the merchandising of new products and the creation of a progressive public image. So much was this the case that some businessmen even came to see the company research laboratory as an attractive investment simply for public relations purposes. As early as the 1920s, Charles Kettering felt it necessary to distinguish his research lab at General Motors from those of companies that created "an elaborate and showy research department" primarily as an advertising ploy and a place to show off to visitors.[4]

General Electric, in 1900, had created the first industrial research laboratory in the United States. Within two decades, some five hundred companies were claiming to possess such laboratories, and by 1931 this number reached sixteen hundred. As Leonard Reich observes in his history of industrial research, business leaders were coming to recognize what GE chemist Arthur Little called "the handwriting on the wall." Industrial markets would soon be dominated by "science-based products and processes." Major companies no longer dared remain dependent on outside sources for new technologies. Recognizing, also, the public relations benefits of identifying themselves with direct contributions to the nation's standard of living, corporate leaders declined to recognize significant distinctions between science and technology, between fundamental and applied research. They emphasized the way in which the scientific discoveries in their labs were quickly converted into new or improved products, often pointing out the democratic implications of this harnessing of science to the needs of all the citizens. Through the corporate research lab, they proclaimed, the findings of science took form immediately in inexpensive products available to all.[5]

For both merchandising and political purposes, corporate leaders came to appreciate the value of a public sense of confidence induced by the promise, now seemingly guaranteed by the institutionalization of technological advancement, of new and better things to come. Almost inevitably, they looked to the opportunities presented by fairs and expositions to dramatize their technological capacities and accomplishments. But they remained of two minds about the importance of wide public access to an understanding of scientific thinking and methodology. Some agreed with philosopher John Dewey that "the future of our civilization depends on the widening spread and deepening hold of the scientific habit of mind," but others perceived, and accepted as inescapable, the growth of an unbridgeable gulf between scientist and layman.[6]

Despite a lingering hopefulness on the part of some business leaders and scientists that their exhibits of science would be truly educational and would inspire incipient scientific "explorers" and inventors, most companies followed their new professional designers toward an acceptance of the intractable credulity of the masses. By the time of the 1939 fair, as Peter J. Kuznick observes, the struggle over the role of the layman had largely been resolved in favor of the "corporate vision," in which nonscientist citizens accepted their role as "relatively passive, though highly entertained," consumers.[7] For that role, an invitation to experience the "atmosphere" of corporate visions of future "worlds" clearly served the visitor more functionally than any effort to explain complex processes.

In tracing both the ideological continuities and the transformations in presentational strategy in corporate exhibits over much of the twentieth century, we focus upon a series of significant moments in the evolution of technological display. From the Parade of Progress and the House of Magic of the 1930s through the Futuramas and Horizons of the 1960s and 1980s, the irrelevance of conventional notions of social or political reform, in a world overseen by the corporate scientist-engineer as pioneer and magician, become apparent.

"Boss" Kettering and the Parade of Progress

In the mid-twentieth century, no single figure so epitomized the increasingly blurred distinctions between science and industry and between engineer and scientist as Charles F. Kettering, chief of the Research Division of the General Motors Corporation. Nor did anyone personify so dramatically the tensions within the general public and among scientists and engineers that were created by the progression toward more complex technologies and scientific theories and a less individualistic, less spontaneous mode of scientific research. Above all, Charles Kettering emerged during the 1930s, the great decade of fairs and expositions, as the corporate leader most ardently involved in fashioning visual displays of popular science and corporate technology.

Kettering was an engineer rather than a scientist. His formal education had gone only as far as a bachelor of arts degree in electrical engineering from Ohio State University. But Kettering was the kind of inventor and hands-on practitioner of applied science that the American public was easily persuaded to enshrine as a scientific hero. Moreover, he loved to perform as a popularizer of science. In 1916, at a conference aboard the SS *Noronic* on the Great Lakes, Kettering presented his fellow engineers with a demonstration of fundamental scientific principles through the "thrilling visual effects" of such "magical" stunts as frying eggs in a skillet resting on a block of ice and conducting high-frequency electricity "right through his body to illuminate a lightbulb held in his hand." According to one of Kettering's fellow research workers, this lecture was "so unique in its scientific viewpoint . . . for that early time" that it opened up "undisclosed vistas" and "was remembered for many years by those who heard it."[8]

General Motors made maximum use of Kettering's instinct for show business and his popularity as a speaker: he seemed never to fail to ignite sparks in his audience with his "salty language," his "quotable aphorisms," and his "rare gift for making science understandable to the layman." In his rising visibility as a spokesman for General Motors, and for industrial research generally, Kettering was also fortunate in gaining the aid of two of the great popularizers of the era, advertising executive Bruce Barton and science writer Paul de Kruif. Barton helped give a somewhat reluctant Kettering national exposure in 1924 with a laudatory essay, full of down-to-earth Kettering stories and examples, in *American Magazine*. De Kruif confirmed Kettering's celebrity status in 1933 with a series of biographical articles in the *Saturday Evening Post*, written in de Kruif's typical gee-whiz, adulatory style. In de Kruif's incessantly "humanized" portrait, Kettering emerged as a "monkey wrench scientist" who disdained higher learning and pure science, preferring to tinker, a more "democratic" mode of research. Through experience in the college of hard knocks, he had managed to "boil [science] down to something . . . deep and simple." By the mid-1930s, Kettering had become America's best-known engineer. Although only an engineer and "old-fashioned inventor," he also became president of the American Association for the Advancement of Science. A contemporary poll ranked him second only to Henry Ford among the "best publicized of all American businessmen." In 1939, *Fortune* would identify Kettering as "science's No. 1 salesman."[9]

During the 1920s, General Motors publicized its research laboratories as part of its marketing efforts and its construction of a corporate image. Kettering contributed to this effort with articles and interviews, lecture-demonstrations, and in the early 1930s, science displays for GM exhibits at auto shows. In 1933 and 1934, at the Century of Progress exposition in Chicago, General Motors made science the central element to regain public confidence in big business and to resist the "regimenting" qualities of the New Deal. In so doing, GM simply appropriated for its own corporate image the theme "Science, Applied to Industry," which fair manager Lenox Lohr had conceived for the exposition as early as 1929 and which the official fair guidebook had captured in the slogan "Science Finds — Industry Applies — Man Conforms."[10]

Noting the rising importance of scientific research to industrial progress, Lohr embarked in 1930 on a campaign to convince the public that scientists were not long-haired men with thick spectacles peering through microscopes in darkened labs but businessmen who worked as part of the "great staffs" of big corporations. As he worked to popularize these unknown "modern silent men" who worked behind the corporate scenes on behalf of a higher standard of living for the public, Lohr also reminded the corporations to keep their explanations of science simple. The fair's mission was to buoy people's optimism about the capacity of American business to restore prosperity through scientific advances. It would be a mistake to "depress them by showing them any things they cannot readily grasp."[11]

Meanwhile, alarmed by the tendencies of the early New Deal, GM President Alfred P. Sloan Jr. concluded that the American business system was threatened by "the thinking of men who believe that progress has ceased, that we must live by dividing up available jobs and accept a lower standard of living." He solicited "nearly three hundred expressions of opinion" from leading corporate scientists and executives on the prospects for the progress of science and industry and publicized their replies at a massive, well-publicized dinner to celebrate the reopening of the GM exhibit at the Century of Progress exposition in 1934. These authorities agreed, a GM press release assured the public, that "amazing things are in store for the world of tomorrow."[12]

Charles Kettering proved himself equal to the challenges set forth by Lohr and Sloan. In fact, it was Kettering who persuaded Sloan to highlight corporate scientific research in his campaign to persuade the public that business could solve the problems of the Depression. Sloan borrowed Kettering's slogan, "The World Isn't Finished Yet," to express the company's optimism about endless scientific and technological advances in the future. Kettering's research laboratory combined simplicity, catchy phrases, and showmanship in demonstrations of scientific equipment at Chicago in 1933. The next year, during the fair's 1934 revival, Kettering expanded, further simplified, and accentuated the research lab's exhibit by building a separate hall within the GM building to display it. In addition to restoring confidence in business, GM concluded, this exhibit would give the company "a priceless identification with scientific progress."[13]

After the fair ended, Kettering and his lieutenants at the Research Division persuaded GM to recapture its extensive financial and technical investment in the research display by putting it on the road as a traveling public relations excursion to small-town America. Thus, in 1936, General Motors launched its Parade of Progress, a caravan of eighteen jumbo, thirty-three-foot-long, streamlined trucks to haul the displays and to serve as exhibit halls; fifteen new-model GM automobiles; and a crew of fifty. Unflinchingly publicized as a "circus of science," the Parade of Progress visited 146 cities, mostly of populations between ten thousand and fifty thousand, during the next two and a half years. It provided tent shows of the wonders of science to Americans in cities too small to have hosted auto shows and that had never experienced exhibits comparable to those on the GM caravan. So gratifying was the popular response—during some local visits, attendance approached or surpassed the official city population—that within six months the corporation developed a "midget" version of the exhibits, carried in one huge van. This show was called Previews of Progress and told the same story of industrial research and material progress through some of the same demonstrations. Focusing on school appearances, it covered communities (averaging ten thousand in population) too small to warrant a three-or-four-day encampment by the Parade of Progress.[14]

Science, translated into a series of magic tricks for the popular audience, was the star of this show. From the beginning, the Parade of Progress was Kettering's "baby." He "worked right along with us," recalled Allen Orth, creator of the Parade's exhibits. Kettering concocted the believe-it-or-not types of snappy phrases that "intrigued the visitors." Kettering also gave the initial lecture-demonstration in Miami that launched the caravan. Souvenir brochures and local newspaper ads proclaimed such attractions as "frozen motion," "music on a light beam," "water boiling and an egg frying on a cold stove," and "the magic eye." All of these phrases sought, as Kettering insisted, to translate science and technology into the "common man's language." Not only was the transformation of a fair exhibit into a traveling road show a thoroughly appropriate tactic for a company that sold physical mobility, the Parade of Progress also promised, through the immediacy of the experience with science that its stage show and visitor-operated exhibits offered to local townsfolk, to reinforce faith in science and technology as the guarantors of a better future and to identify General Motors with that faith.[15]

Kettering could preach that faith with confidence and imbue his exhibits and demonstrations with its motifs because he had thoroughly assimilated and then helped popularize a dynamic theory of history. Resting basically on a confidence in the beneficent and democratic qualities of a free market economy, this theory enjoyed the support of most business leaders. Like other business spokesmen, Kettering rejected all conventional utopian visions as inviting the stagnation and lack of stimulation inherent in any static ideal. His faith in the unlimited possibilities of new inventions led him to place his faith in the vision of a never-finished society of inspiring challenges, increasing material comforts, and ever-expanding markets.[16]

Major elements of this perpetual motion theory were implicit in Henry

Ford's championship of mass production and the five-dollar-a-day wage. Efficient mass production would lower the cost of goods; high wages would turn workers into consumers, thus reinforcing the need for more production and providing everyone with a higher standard of living. Barton and his advertising colleagues added the element of incentives: "You go into a savage tribe and what do you find? You find men who have no wants. . . . So a savage tribe continues for a thousand years and there will be no change. . . . But suppose that out of an airplane an advertising man dropped into that tribe and brought with him pictures of red neckties and tan shoes and underwear and new hats and automobiles. . . . Instantly there would begin in that tribe a transformation. Wants would be kindled . . . and in obedience to them even a savage is willing to abandon his life of leisure and voluntarily enlist himself in servitude to the creation of a civilization."[17]

In *The New Necessity*, a promotional book that Kettering coauthored with Allen Orth for General Motors and the Century of Progress exposition in 1933–34, Kettering enlarged upon that theory of progress. Kettering had won the plaudits of the editor of *Advertising and Selling* in 1929 by "openly and without apology" proclaiming that "the main function of industry's research departments is to make people dissatisfied with what they already have." While "the bolsheviki smash with an axe," the editor observed, Kettering's research smashed complacency "with an alternative." As the Depression continued, Kettering denounced the old regime of thrift and promised deliverance from hard times through an intensified cultivation of people's dissatisfaction. Wherever needs and wants were in balance, he argued, civilization was static and "unhealthy." When people began to want things they didn't "need," they became "more alert mentally, more willing to work, more willing to do the unusual." The lesson of the automobile had been that "the world wants new things, new ideas, new products, new opportunities." Science, which was now carried out more systematically and efficiently by "groups of scientists" working "along certain well-organized paths" in corporate laboratories "beyond the imaginings of the wildest scientific dreamer" of the past, ensured that a steady flow of new inventions would keep the society dynamic and endlessly fulfilling.[18]

Kettering's tactics, in parallel with exhibits of corporate science and technology, replicated this perpetual motion, endless frontiers theory. Mystifying and awe-inspiring scientific tricks suggested the infinite possibilities of science, the new frontier that would afford future generations the challenges and opportunities once derived from the geographical frontier. Through its capacity to create new products, new needs, and new jobs, Kettering argued,

GM offered people a vision of new frontiers, in contrast to the New Deal's acceptance of stagnation. The lecturers and exhibits of the Parade of Progress never failed to point out that General Motors would quickly seize upon each discovery demonstrated in their display of the "magic of modern science" to improve Americans' standard of living.[19]

Kettering had high hopes for the efficacy of popularization. He fully expected audiences to achieve an understanding of science through tricks and demonstrations that "reduced complicated ideas and projects down to simple, easy-to-grasp things." But the experiences of demonstration lecturers, at the Parade of Progress and at other corporate exhibitions and displays, did not always confirm that expectation. The tricks held the attention of the crowds, even "dazzled" them. The audience in Greensboro, North Carolina, according to a local newspaper, was "literally taken off its feet so unusual were the demonstrations." But it is dubious whether the emphasis on wonders and magic was conducive to the scientific cast of mind that some reform-minded scientists and engineers saw as essential to the progress of democracy and civilization. Instead of introducing the audience to a scientific mode of thought, corporate exhibitors played to the audience's penchant for magic and then reassured the audience that the corporate scientists who possessed the knowledge of such magic were dedicated to making it serve the public in practical ways.[20]

GM's view of history played a major role in instilling this point of view. As its director of public relations put it, the Parade of Progress and the midget caravan "constantly contrast the old and the new and project our audiences into the future." Historical exhibits contrasted a home of 1900 with one of "today," invited chuckles at an antiquated 1902 Oldsmobile, and posed models of modern service stations and automobiles against diorama backdrops depicting a village smithy and "Grandpa's ol' Dobbin parked before the corner saloon of bygone days." Each suggested the inevitability of technological progress. This view of history served to provoke a snicker of amusement at yesterday's inadequacies. The people of earlier decades had also snickered — but at precisely the wrong things. They had laughed at the new technologies and had even "violently opposed" them. Part of their derision had stemmed from their mistaken assumption that such new contraptions as the automobile would be nothing but playthings for the rich. In the face of such blind conservatism, which Kettering never missed an opportunity of condemning, scientists and engineers — with their can-do outlook and their refusal to accept the notion of the impossible — had brought progress to a complacent and unimaginative populace and had ensured the democratic impact of new technologies. A scientific elite would provide the dynamic element needed to overcome public inertia and ensure an ever-changing, open-ended future.[21]

For all of Kettering's stance as the champion of progress, modernization, planned obsolescence, and the certainty of advance through organized, systematic research, he still represented more vividly than any of his fellow leaders in corporate research a transitional figure with strong ties to a bygone age. In fact, it may well have been Kettering's self-cultivated image as a "screwdriver and pliers man" that made him an acceptable popular image of corporate science. Kettering reminded observers of "an old-time country doctor," one newspaper observed, and he liked to be called a research worker rather than a scientist. He loved to make fun of the "experts" and to exult in accomplishing what they claimed was scientifically impossible. As a down-to-earth, small-town boy, Kettering had learned much of his science, one of his colleagues relates, by conducting experiments during his free hours at a small-town drugstore. He took satisfaction in relating how he had developed the self-starter "in a barn with only a lathe and a drill press as machinery" and had continued to "prefer the intimacy and equality of the Barn Gang days." [22]

Thus Kettering, head of one of the world's largest industrial laboratories, employed the rhetoric of the "shop culture," of the independent engineer claiming to represent real science (the down-to-earth, experimental, applied kind) and avoiding relinquishing his self-image as "just a plain old inventor." So pervasive in American culture in the early twentieth century was the metaphor of the scientist as pioneer and explorer that Kettering—and other industrial scientists and engineers—easily adopted this romantic image, with many of its individualistic implications. DuPont, which had publicized the chemical engineer in the early 1920s through the image of the lone researcher gazing upward—literally at his test tube but symbolically into the future—continued to romanticize the pioneer image by employing such titles as "Blazing the Trail to New Frontiers through Chemistry" in its 1939–40 World's Fair exhibit. Owen D. Young of General Electric celebrated GE research workers as "explorers [who] set sail for unknown lands," and Kettering talked steadily about "the spirit of pioneering" in applied science and the promise of "frontiers unlimited." [23]

Although the reality of large corporate laboratories was scientific teamwork and systematic, collective endeavors, their preferred image of the scientist was that of Edison, "a hero ever forging ahead, ever seeking to explore new lands." In 1929 Maurice Holland even gave his portrait of the heads of corporate research laboratories the title *Industrial Explorers* and characterized his subjects as trail blazers and master explorers who worked on the "frontiers of science." Unquestionably, it aided the popularization of industrial research to give it this humanizing, individualized face. Moreover, the image of the lone explorer and pioneer served, as John William Ward illustrates in his essay on Charles Lindbergh, to negotiate an uncomfortable transition to a more complex and bureaucratic society. The implication of the pioneer—that anyone with courage and vision could have done it—also suggested continuing democratic access to the knowledge and power that science promised.[24]

The pioneer metaphor was already anachronistic during an era in which Kettering was bragging about having put science "on a production basis" and was warning private inventors that "the average man" was totally out of his class in trying to compete with "the carefully organized effort of trained specialists." In 1932, in the process of explaining the necessity of corporate research laboratories to "assure a systematic change-making," Kettering published statistics likely to unnerve even the most intrepid amateur inventor. Fewer than one out of five thousand inventions submitted to manufacturers by independent inventors, he contended, contained any merit.[25]

Still, the Parade of Progress publicized its display of science in ways that preserved the individualistic, amateur-inventor image. In 1938, GM developed a follow-up advertisement for local newspapers that replicated the heroic inventor of old in the image of a lone boy who turned to his home chemistry set or tool kit to make his own scientific explorations. Entitled "We hope we set a boy to dreaming," this much-lauded advertisement linked General Motors' fusion of science and industry to the "real hope for continued progress" that lay in "the spirit of individual initiative." The protagonist, a boy of about twelve or thirteen years, surrounded by test tubes, microscope, and beakers, gazes idealistically upward, cheek in hand, with that same quality of "self-absorption that excludes the beholder" that David Nye identifies as typical of the portraits of GE inventor-scientists by company photographers. The simple key to creating the ad, the agency copywriter observed, had been to express "the sincerity of feeling which existed among those who were putting on the 'Parade of Progress.'" It was precisely this "sincerity," perhaps better characterized as sentimentalism, that enabled leaders of corporate science to retain the self-image of the lone frontiersman while they were actually pioneering the advance of systematic collective research.[26]

The boy-with-a-dream figure, anachronistic as it may have been, did suggest a solution to the most disquieting ramification of the perpetual motion theory of scientific progress. Most of the businessmen who championed that theory adhered to a hard work, producer ethic system of personal values. Nor was there anything about a commitment to science to suggest sympathy with consumeristic indulgence. John Burnham observes that proponents of science in the nineteenth century had "linked belief in natural order with a lifestyle emphasizing service and self-denial," and Marcel LaFollette, in her study of the popular images of scientists between 1910 and 1955, notes how commonly popular magazines defined scientists' "eagerness and willingness to work far beyond the normal work-day" as a central characteristic that set them apart from "ordinary people."[27]

These were men who liked work, valued an image of austerity, and described science in sweat-of-the-brow, workbench terms. Could they look favorably upon a future in which the benefits of science would lead the nation into an overly comfortable self-indulgence? And from where, in such a society, would subsequent generations of ambitious, self-sacrificing pioneers of science come? The largely unspoken answer to this dilemma emerged in the conception of a radically bifurcated mass society. In this world, the qualities that had once defined the male gender at large would be revitalized each generation only among a small elite drawn from the ranks of those special boys whom the displays of scientific prowess had set to dreaming.

Thus the biographies, reminiscences, and anecdotes of their venerated "Boss Ket" by Kettering's research aides and disciples shaped their hero in the image of the dirty hands, work-obsessed figure so dominant in popular depictions of the scientist and engineer. Kettering was a man, the anecdotes proclaimed, who would happily abandon any form of leisure, any cultural event, any formal gathering in pursuit of a technological breakthrough. He would crawl underneath an engine in his dress suit, if need be. T. A. Boyd, Kettering's research associate and biographer, seemed to revel in recounting all those occasions when Kettering deserted his wife — both emotionally and physically — to meditate on an engineering challenge or join the boys in the lab. Rosamond Young, another Kettering biographer, recounts almost with relish the many slights and inconveniences his understanding wife had to suffer because of his higher allegiance to science. "Poor Olive! She lived a lonely life," Young concludes. "But with Charles for a husband that is the way it had to be."[28]

And so it did, in the dominant, highly genderized model of a society moving toward creating material fulfillment through science. Science connoted the harsh, male world of "ruthless rationality and cerebral dealings"; scientists' "neglect of domestic relations, long periods away from home" suggest conventional male patterns of behavior as well as their devotion to their work.[29] Women — like Olive, who "loved the shopping and sightseeing" — played the necessary role of beneficiary of scientific advance, repaying through gratitude and support the scientists who enabled them to enjoy an increasingly debilitating consumption.[30] They were also the consumers of that commodity that the true scientist and engineer most feared and despised: leisure. As technological and scientific progress engendered vaster quantities and varieties of goods, most men were inexorably drawn into the feminized ranks of consumers. But as long as an elite cadre of scientists and engineers could find their (more austere and sacrificial) fulfillment in pioneering the frontiers of science and in assisting the populace to reap the material blessings of science, a dynamic vision of endless progress could be maintained.[31]

Such progress-without-enervation depended upon one essential social mechanism — a process set forward in the ad "We hope we set a boy to dreaming" and implied again and again in the exhibits of corporate research at the 1939-40 New York World's Fair. A radical bifurcation of mass society would ensure that most citizens gained their fulfillment through consumption, enjoying an ever-higher standard of living. Meanwhile, a small elite, steadily replenished from the ranks of serious, dreaming boys (dreaming girls during this era sat by hope chests, not microscopes), would explore the frontiers of science and technology and, uncorrupted by luxury and leisure, enjoy the more rarified fulfillment of a life of challenges. Every demonstration at the Parade of Progress stage show served to convey this image of service to the feminized millions by a dedicated, pioneering elite. The tacit promise of the Parade of Progress was not the reformist hope of building a better world through universal rationality and the popular adoption of the scientific method. Rather, it was: trust us and we'll do it all for you.

The House of Magic

At Chicago's Century of Progress exposition of 1933-34, where General Motors introduced to a national audience displays of the wonders of science that would characterize the Parade of Progress, another exhibit titillated the public curiosity about scientific wonders even more obtrusively by labeling itself the House of Magic. This invitation to peek into the occult was not located in the amusement zone; it was the scientific exhibit of the Research Laboratory of the General Electric Company, which, in its mass media advertisements, had already used the nickname House of Magic for its laboratory.

Since its founding, the GE laboratory had performed the auxiliary purpose of serving as a "company showplace, where prominent visitors . . . were taken to be shown that General Electric put a strong emphasis on scientific research." In 1923 the use of the laboratory as a major element in corporate publicity was extended beyond scientists and industrialists to include the general public, as the new GE president, Gerard Swope, embarked upon a new merchandising strategy. In a campaign to instill an "electrical consciousness" in the public, Swope sought to expand sales for large electrical appliances by extolling the use of electricity. He also looked to associate the GE monogram with dedication to science to establish "consumer acceptance" for new GE ventures in consumer appliances. Surveys by GE's advertising agency in 1924 and 1926 to track the effects of the multiyear "electrical consciousness" campaign revealed that people were more aware of GE's activities in research than of any other corporation's.[32]

During the 1920s, GE's publicity chief, Clyde Waggoner, also sought to heighten the lab's reputation by publicizing such company luminaries as Willis Whitney, Irving Langmuir, and Charles Steinmetz. The Barton, Durstine, and Osborn advertising agency spotlighted Steinmetz in GE's institutional ads, and agency head Bruce Barton constantly publicized Thomas Edison and worked to keep his name linked with GE. But the exploit that did the most to implant GE research in the popular consciousness was carried out by a flamboyant, fast-talking journalist and adventurer whose gregarious style and romantic adventures as a war correspondent provided the atmosphere for introducing the mysteries of GE research to the radio public. Beginning in November 1929, during the GE symphony hour, correspondent Floyd Gibbons presented ten-minute talks to publicize the GE laboratory. It was he who dubbed the Schenectady facility the House of Magic. Gibbons explained that his ignorance had qualified him for the task of making corporate research intelligible to the public and confessed that it took him "some time to boil this mess of scientific porridge down to the consistency where I could digest it." Having done that, he concluded that the story of the "wizards" at the GE lab was "the weirdest tale of hocus-pocus I ever heard." For the next two years he rhapsodized about GE magic to a radio following that reportedly included "a vast audience of women . . . many of whom frankly admitted that they never expected to become interested in such subjects."[33]

Fortune magazine later characterized Gibbons' "breathless" accounts of GE wizardry as the high point in the glorification of corporate research. Gibbons, himself, observed that GE scientists had initially "frowned" on his use of the term House of Magic but had soon come to like it. Correspondence among GE officers reveals a more ambivalent, and sometimes contentious, attitude. The head of the GE laboratory, Willis Whitney, recognized the danger in the image of magic. According to historian George Wise, Whitney felt uncomfortable with what was being said about "houses of magic and miracles." Whitney told a journalist in 1930 that he "personally would like to stop it." He took pride in being a worker—but not of miracles. Nor was Whitney alone in his discomfort with the slogan. Lawrence Hawkins recalls that Gibbons' phrase "caused considerable mental anguish to some of the research staff." In 1930, Charles Wilson advised GE president, Gerard Swope, that Gibbons' broadcasts had made the appellation an "accomplished fact" and even a subject for joking allusions in a recent movie. Despite his evident discomfort, Wilson concluded that use of the phrase might have "gone too far to be killed." Bruce Barton, one of GE's advertising agents, forwarded a letter to Gerard Swope from an advertising journalist who spoke of a "cheapening" of the GE image by too much hokum about the laboratory.[34]

Company advertising and publicity men, however, loved the nickname. They heralded a new GE radio in 1930 with ads headlined "Out of the House of Magic" and included a lengthy, conversational account of the laboratory by Floyd Gibbons, who in recounting the story of GE radios used the words *wizardry* or *wizards* four times to characterize GE scientists. The House of Magic slogan had "tremendous commercial value," observed GE advertising man Martin Rice. Once the resistance of the research laboratory had been largely overcome, the company satisfied its conscience by employing the circumlocution, "Our Friends call it the 'House of Magic.'"[35]

At the Century of Progress exposition, GE dropped the cumbersome phrasing and simply labeled its exhibit House of Magic. GE publicity releases and the *GE Monogram*, a company magazine for its managers, called the 1933 display "a thrilling presentation of electrical wonders" in "thirty-minute shows consisting of six or more short acts." Stroboscopes, oscillographs, and high-frequency coils would be the "star performers"; astounding tricks and marvels would occur with "the pushing of a tiny button" or the "wave of the scientist-magician's wand." According to GE and Century of Progress publicity releases, the House of Magic would put "electricity wizardy" on display through a "bag of tricks," including a gun that would "reverse the order of the 'wild west'"—shooting "on the lights instead of shooting them out." However, a Ford Motor Company report on the 1934 Chicago exposition compared the GE exhibit very unfavorably with one by the Massachusetts Institute of Technology, noting that the House of Magic "plays with the stroboscope and makes it entertaining, [while] the lecturer in the MIT exhibit really illustrates its principle." At General Motors, furthermore, a 1932 memo assured its public relations committee of the scientific seriousness of a Paul de Kruif magazine serial on Charles Kettering, declaring: "This will be no 'House of Magic' stunt." In 1936, a GE report described how its technical brochures avoided a " 'House of Magic' idea" by "capitalizing on research information in a dignified way." [36]

These divergent views of the House of Magic image reflect the dilemma of the popularization of corporate science. Nearly all segments of the scientific community in the early twentieth century had come to recognize that the popularization of science was necessary to ensure public and, especially, financial support. Many scientists also felt a missionary instinct to infuse others with their own enthusiasm and understanding. By the 1920s, moreover, an increasing number of corporations in science-based industries saw a vital link between their future marketing success and their public reputation (the public's appreciation of their expertise in science and technology). Those who had smarted from the lash of the muckrakers hoped particularly to defend their bigness by bringing the public into an understanding of their business operations and services, an understanding that they believed required some familiarity with, or at least an appreciation of, their technology and the scientific basis of their processes of manufacture.

Rather like a Christian missionary who assents to aspects of a native "cult" to gain the natives' conversion, corporate representatives and exhibit designers often appealed to the popular preference for embracing science and technology in terms of the fantastic and the miraculous. Scientists with a primary commitment to the teaching of science, and those concerned about the fate of a political democracy in the absence of a popular propensity for rational analysis, might hold out for a sterner, more didactic approach to popularization, one that stressed the schooling of laymen in scientific thinking and the scientific method. But corporations proved increasingly disinclined to insist on such a rigorous and austere approach to the public, particularly in the atmosphere of fairs and expositions.

The object of science-based corporations like General Electric, in exhibits that served the purpose of advertising and public relations, was not to make demands upon people but to demonstrate how the corporations served people. If gaining true understanding proved difficult, these corporations were easily persuaded to settle for appreciation. [37] The apparent popular desire for an Aladdin's lamp interpretation of science and technology—with the interpretation of the machine focused, in Dorothy Nelkin's phrase, "in terms of what it did, not of how it worked," and with the increasing emphasis on, in John Burnham's words, "portraying the results, rather than the ideas, of science"—coincided perfectly with the corporate motive to present itself as the benevolent and competent genie in the bottle, eager to make its wizardry of use to the public. The

N. W. Ayer and Son agency, advertising agent in the late 1920s and early 1930s for American Telegraph and Telephone and Ford, urged the public to view new consumer goods, from automobiles and flashlights to cold cream and movies, as "everyday magic." The most prevalent "image theme" in magazine articles on science, Marcel LaFollette observes, communicated the message that "science is magic and the scientist is a wizard." [38]

In such an atmosphere, and given the public relations objectives of the exhibits, it was hard for General Motors or General Electric—or any other science-based corporation—not to take advantage of the visual dimension of exhibits to put on a "magic show." Thus, GE employed a professional magician to perform at its House of Magic exhibits at the various 1930s fairs, and Harry Manning, lecturer for GM's Parade of Progress and Previews of Progress caravans, presented the "talking flashlight" in a symphony of "three magic movements—see, whee and gee whiz." GE turned its House of Magic into a traveling show and, according to an associate of designer Walter Dorwin Teague, resolved to "reproduce the House of Magic—in even more magical form" for the New York World's Fair. [39]

The Future, in Animation and Automation

The New York World's Fair of 1939 erected the largest, most visible stage of pre–World War II America, for the display of corporate science and technology. By this time, large corporations, like AT&T, General Electric, DuPont, Standard Oil, General Motors, Ford, U.S. Steel, and a score of others, had accumulated considerable experience in "industrial showmanship" and had also turned to professional industrial designers, such as Walter Dorwin Teague, Norman Bel Geddes, and Raymond Loewy, for expert guidance. [40] Almost everybody was in the magic business now. AT&T and Westinghouse had their audiences talking with robots, and AT&T magically enabled "auditioners" among their visitors to hear their own voices emerge from the mouths of dummies in a stage presentation. The main purposes of its "intriguing demonstrations," AT&T noted, were "to enlist participation by visitors" and to "employ scientific novelty of such a nature as to be easily understood by the general public." General Motors featured "feats of scientific magic" in its Casino of Science show. GE's House of Magic ranked first among exhibits that "mystified science through dramatic effects," in Peter Kuznick's phrase. *Business Week* characterized the GE presentation as "pure science . . . served to the public as unalloyed entertainment." [41]

Chrysler Motors advertised its exhibit as a five-star show. Called Sixty Minutes of Magic, the Chrysler extravaganza featured a simulated ride in a rocket to London, a "frozen forest," and "engineering wonders." Chrysler's Magic Car opened and closed its doors, blinked its lights, and answered questions from the audience. To capture and hold an easily bored audience, Chrysler's public relations firm explained, the company had turned first to the spectacle to gain attention and then adhered to "the newer concept of what an industrial exhibit should do." It should impress the audience with what the company and its products could do for the consumer, not focus "on the complicated processes" of production. Thus, the essential purpose of even the section of the exhibit with the most "informative design," the display describing the "actual tests of the materials and [Chrysler's manufacturing] processes and operations," was to "give visitors a dramatic, compelling experience." [42]

Even those firms most wedded to an older notion—that fair exhibits should foster a scientific and engineering frame of mind by demonstrating the processes of manufacture—found it impossible to resist the stampede toward audience-pleasing showmanship. At the 1939 New York and San Francisco fairs, DuPont still presented lectures on the wonders of the chemical transformation of raw products, as it had at the 1936 Texas Centennial in Dallas. But it radically simplified its presentations on the advice of its professional designer, Walter Dorwin Teague, who recommended the "complete elimination of all technical details not familiar to the public, with the whole complex process reduced to a few dramatic steps." It also added a marionette show and demonstrations in which, "by a trick of science," live models fashionably clothed in DuPont materials were "made to appear from an apparently empty test tube of 'Lucite' plastic." The company's Miss Chemistry, the DuPont fair brochure of 1940 proclaimed, magically emerged "as if she herself, in addition to her costume and accessories, had been created by chemistry." [43]

The Ford Motor Company, which had presented an operating assembly line at a San Francisco fair in 1915 and elaborate displays of essential steps in the production process at the Chicago fair in 1934, now "educated" visitors in the processes of its "cycle of production" through a massive, thirty-foot-high, rotating cyclorama of animated figures. The audience observed the transformation of raw materials into automobiles by watching the animated antics of eighty-seven groups of carved figures of men and animals with "humorous details . . . added through plastic noses, eyes, horns and tails." For the more sophisticated visitors, Ford offered a huge, three-dimensional, "activated mural" in which the abstracted forms of a Ford V-8 motor—including pumping pistons and "whirling gears" against a background of scientific symbols, formulas, and images of the Ford factories—conveyed the message that "mass production is made possible by modern technology, and that modern technology is based on the principles of pure science." Ford's radio spokesman, W. J. Cameron, found "something mystical" about it, and the fair's director of science and education, according to a Ford press release, paid homage to it as "an altar piece of science."[44]

During 1940, visitors to the Ford building, after their exposure to the "symbolism . . . intended to excite the imagination" of the "altar piece of science" and their "instruction" in processes through the animated antics of the figures in the "cycle of production," could then enjoy a live musical comedy entitled "A Thousand Times Neigh," through which they could place current automobile technology in a historical context. The show, which Walter Teague described as having "a fantastic, Walt Disney quality," heaped ridicule on the skeptics of inevitable technological progress by recounting the history of transportation since the era of barbarian horsemen. In one scene, the script called for male dancers to celebrate the invention of the wheel, as follows: "men who are freed from the labor of moving the stone turn cartwheels about the stage (play the whole pantomime for comedy) . . . and the entire group begins a dance patterned on circular motion." This drama was presented through the eyes of Old Dobbin, a horse who lightheartedly acknowledged his own obsolescence and encouraged visitors to view applied science as carrying them forward from the benighted past in a continuing and inevitable trajectory into a future of ease and abundance.[45]

For all of the entertainment afforded by magic tricks, animated scenes of production, and musical comedy versions of the history of technology, the indisputable hit show of the 1939 New York World's Fair was GM's Futurama. With its greater audience participation and total, controlled environment, Futurama foreshadowed not only the America of 1960, which it ostensibly portrayed, but also the presentational strategies of future corporate displays of technology. Created by the erstwhile prodigy of spectacular theatrical design, Norman Bel Geddes, Futurama physically moved visitors through its illusionary world, inducing them not only to see the future of the nation as General Motors envisioned it but, in a sense, to experience it for themselves. Bel Geddes constructed a small-scale model of a vast American landscape, complete with more than two million buildings, fifty thousand cars (ten thousand of them in motion), airports, suspension bridges, amusement parks, an elaborate system of freeways, and a planned city of the future.[46]

Spectators traveled over this landscape in banks of moving chairs at a height and speed simulating cruising in an airplane at an altitude of about five hundred feet. Individual loudspeakers in the chairs educated the viewers on the principles of design, lifestyle, politics, and sociology required for the realization of this scientifically fashioned world of the future. As a sensational climax to this display, Bel Geddes brought Futurama riders down close to a particular intersection of the future city of 1960 to view the design for separating the levels of pedestrian and automotive traffic. Then, as the tour ended, spectators' chairs were suddenly "swung about" in a semicircle to deposit them in a full-scale model of the smaller-scale intersection they had just viewed from above. The spectator, in Bel Geddes words, "can scarcely believe his eyes. . . . He gets out of his chair and becomes part of the crowd," walking about the streets of the "future" and entering the buildings on the corners of the intersection to view further GM displays. Given this new advance in presentational strategy — from "how we do it" to "imagine yourself already there in our new world" — it seemed a permissible flourish to offer each departing visitor a pin that read "I have seen the future."[47]

Futurama was an overwhelming success. Even more momentous was its impact on the evolution of corporate display. Futurama prefigured many of the themes and techniques with which major businesses, over the next half-century, would portray science and technology as attributes of their corporate identities. Visitors to Futurama witnessed not simply products but an entire future consumer realm reshaped by science, technology, and free enterprise. On a clear day, Futurama suggested, you could see not only General Motors but the world as GM imagined it.

Dioramas on a less ambitious scale had appeared at world fairs since 1893 and had attained prominence by 1933. But Futurama was not simiply a scale model of superhighways and skyscrapers. It represented a world transformed: a landscape (ideological as well topographical) shaped by "progress," a force apparently fueled by the remarkable similarities between corporate growth objectives and scientific research and development. Futurama was quite simply the most elaborate embodiment of corporate futurism ever devised. Its dioramas were at once fantastic and oddly familiar, for it brought together the diverse elements of industrialized America's technological optimism in all its totalizing, streamlined, market-driven, nature-conquering, corporate designed, government subsidized glory. Pehaps the most significant attribute of this mode of technological display was its prominence in mass culture. Although the world envisioned in Futurama rested on a comparatively narrow range of assumptions regarding the social applications of science, few voices challenged its claim of inevitability. Modernity itself had become another design project for company engineers.

Carousels of Progress: The 1964 Fair

Despite the sweeping scientific, corporate, and social changes in the quarter century following 1939, the spirit of Futurama was readily discernible at the 1964–65 New York World's Fair. Contructed on the same Flushing Meadows site as its 1939 predecessor, the 1964 fair followed essentially the same layout and featured many of the same corporate participants.[48] Both fairs highlighted the social role of science and technology. Yet the differences between them revealed a broad shift in presentational strategies. The 1939 fair's effort to balance explanatory depictions with showmanship had given way, by 1964, to fables of progress, special effects, and magical invocations of technological wonders.

For the 1939 fair, corporations like General Motors, General Electric, AT&T, and DuPont had focused much of their public relations on celebrating the useful applications of their research labs. Since World War II, both the public relations offices and the research and development labs of these companies had grown considerably; but they had grown in different directions. In 1939, corporate managers still sought to bridge the gap between production methods and display techniques; by 1964, manufacturing and public relations were typically managed by completely separate departments. For the research labs, both the complexity of their work and the secrecy surrounding it had become much more prevalent. In the wake of the Manhattan Project, the press and corporate publicists alike characterized science and technology as less comprehensible than ever but capable of increasingly fantastic feats. Even if a new generation of Boss Ketterings had emerged to educate the public, there would have been much they could not say.

If science had become less demonstrable, the image of science was more inviting than ever to the growing ranks of publicists and advertisers. *Public Relations News* called the 1964 Fair "one of the most ambitious public relations programs of our time," noting that "in contrast to the 1939 New York World's Fair, where very little attention was paid to any aspect of PR except publicity, *every policy is carefully evaluated in advance in the light of its PR implications.*" Two public relations firms coordinated the fair's activities; the head of the fair's executive committee was a professional PR counsel, as were a number of members of the fair's board of directors; and by 1964, most of the major exhibiting corporations had extensive PR staffs of their own. What had been a growing concern in 1939 had become an institutionalized bureaucracy by 1964.[49]

Both the scope and the aim of publicity had shifted. As publicity and exhibit design became increasingly the tasks of professional publicists and less the concern of corporate scientists and engineers, depictions of technical processes gave way to invocations of the social attributes that corporations and consumers alike were sure to enjoy. It was as if the transfer of products from producer to consumer bestowed magical qualities on both; and those qualities, in turn, were equated with science and technology.

The appeal of a scientific image as a corporate attribute was addressed in the fair's sales pitch to potential exhibitors. Advertisements aimed at the business community portrayed corporate image and national leadership as inextricable attributes of progress, which in turn was driven by science and technology. A 1962 advertisement in *Advertising Age* described the upcoming fair as "magnificent spectacle . . . significant marketplace!" The fair promised to combine "corporate prestige" ("a world stage on which to present your company's scope and stature") with "national image" ("our industries must present our claim to leadership"). Fair president, Robert Moses, proclaimed that this would be an "Olympics of Progress . . . an endless parade of the wonders of mankind." [50]

Promoting the fair to the general public also meant stressing the "wonders of science" theme. A radio commercial prepared by the J. Walter Thompson agency asked, "Want to see a machine that chews up jungles and spits our highways . . . meet a dinosaur . . . see a controlled nuclear explosion . . . make a trip to the moon . . . see the most fabulous tree in the world (its leaves are *money*)?" Another radio commercial announced that "sidewalks will move— highways will be automated. There will be men on the moon and the King James version of the Bible will be obtainable on a computer card. . . . This is just part of the mad, mad world of tomorrow that will be revealed at the New York World's Fair." [51]

If some of the 1939 fair's industrial exhibitors felt torn between educational and entertainment values, the 1964 fair appeared to have tipped the scales toward entertainment. This fair, David Brinkley predicted, would be "a fair rather than a trade show"; in place of "steely phalanxes of turbines and drill presses" that characterized some previous fairs, 1964 visitors would find "a show of what is new and interesting and entertaining in a country that no longer has to prove it knows how to make machines." Brinkley also cited *Variety*'s observation that "those who predict a thud from the 'industrial' side of the fairgrounds are not reckoning with the fact that since the last fair a whole new breed of pro showmen have emerged from the mercantile world." [52]

These "pro showmen" shared none of the misgivings over magic that had been voiced by Willis Whitney and others in the previous corporate generation. "People go to a World's Fair because they are seeking excitement, and that is the only reason they go," claimed J. E. ("Jiggs") Weldy, manager of sales and advertising services for General Electric. At a conference of his fellow 1964 fair exhibitors, he cautioned that "with entertainment you can couple a little bit of education, but not very much, because people don't go to World's Fairs to study."[53] Exhibit designer James Gardner concurred: "If your aim is to interest an exhibition crowd," he told the same gathering, "you are in show business." The "great thing," he explained, was that "people go to these shows just as a child goes to the store to see Father Christmas. They are full of anticipation and excitement. . . . They are psychologically ready for you to influence them." By implementing "controlled circulation" of the crowd, he claimed, exhibitors could cinematically influence their audience: "They are hypnotized. They forget the thing three frames back, but you can lead them right along a line and work on their emotions."[54]

The show business tendencies that emerged in the 1939 fair became dominant in the 1964 fair. Both fairs, for example, looked for entertaining ways to exhibit automatic milking machines. In 1939, the Borden exhibit featured the rotolactor, a revolving platform on which real cows stood as they were being mechanically milked. In 1964, the Electric Power and Light Building included a mechanical cow that "sings about how nice it is on cold mornings to do business with an electric milker instead of a farmer's cold hands."[55]

The 1964 fair's sponsors remained firm in their declarations of faith in progress and in their efforts to link corporate science with national destiny. Faith in progress, however, had become more difficult to sell since 1939. The fair's PR offices struggled to offset the apparent indifference expressed in some quarters of the press corps to the fair's exuberance. Part of the difficulty stemmed from the overbearing personality and inflexibility of the fair's director, Robert Moses.[56]

Exhibitors and publicists, however, also had to confront the irony that, at a time of relative abundance, the idea of progress was not the unqualified assumption that it had been during the Depression. Now that the capacity for great technology seemed within reach, at least some commentators were less sure that that was what the nation should be trying to achieve. Civil rights demonstrations at the fair site reminded the public that technology's cornucopia had not reached everyone, and the Cold War and the arms race cast a shadow over the fair's wonders of technology. Contrasting the need for fallout shelters with the optimism of fair publicity, radio commentator Dorothy Kilgallen complained of the "inconsistency of men in high places": "The Governor [Nelson Rockefeller] and the President [Kennedy] think we are going to be devastated by bombs, but if we have [fallout] shelters maybe a certain number will be saved. On the other hand, Robert Moses thinks everything is coming up roses, or everything's coming up Moses, and we're going to have a World's Fair to celebrate the inevitable sweep of progress."[57]

The fair itself included similar mixed signals regarding the direction of technological progress. In the Hall of Science, the Atomic Energy Commission sponsored Atomsville, U.S.A., an interactive display inviting children to "explore the concept of the atom." Yet the Underground House, built by former chemical warfare expert Jay Swayze, was essentially a luxury fallout shelter with "dial-a-view" simulated picture windows.[58] Even the early draft of the fair's theme—a shrinking world in an expanding universe—hinted at a certain ambivalence toward the future. For the most part, however, the promoters of the 1964 fair struggled to produce an updated rendition of 1939's hymn to progress, increasing the volume wherever the lyrics seemed weak.

Few companies were as adept as General Electric at equating their products with progress; according to a GE advertising campaign, "Progress is our most important product." GE's 1964 pavilion went well beyond the House of Magic exhibits of 1933 and 1939; GE's Progressland presented GE as an active contributor to the full sweep of technological progress in American culture. GE radio ads encouraged listeners to attend the fair, with only passing mention of the GE pavilion's attractions: "an actual nuclear experiment, a 4-act play designed by Walt Disney and a galaxy of science and engineering." ("It's wonderful to work for a company like General Electric," the fair's special exhibits director wrote to a GE executive, "where you can afford to do radio spots practically without mentioning the sponsor's product!").[59]

Progressland exhibits embraced a range of styles. The Electric Power Generation Exhibit and the Steel Mill Exhibit came as close to educational demonstrations as anything at the fair. The much-publicized fusion demonstration combined circus-of-science hype with claims of scientific credibility. Genuine "General Electric scientists" were on hand, making this one of the few exhibits anywhere at the fair to include technicians as demonstrators; their instruments provided the audience with "bonafide evidence that true nuclear fusion was accomplished as they watched." Elsewhere in the pavilion, the Skydome Spectacular borrowed more directly from the House of Magic's special effects approach: viewers watched a huge overhead concave screen as film projections dramatized "the story of man's historic search for energy" from cavemen to the atom.[60]

GE's most popular exhibit was the Carousel of Progress, designed by Walt Disney and featuring his new "audio-animatronic" mechanical figures. The Carousel's audience revolved about an immobile central stage that had been divided into slices of progress. Each section depicted the same all-American nuclear family enjoying the fruits of consumer culture at a different moment in the past century: first, gas lamps and telephones; then, electric lights and phonographs; then, refrigerators, radio, television, and electric dishwashers. The idea was not entirely new: in the 1930s, GM's traveling Parade of Progress contrasted "homes of yesterday and today," and at the 1939 fair, Westinghouse's Battle of the Centuries exhibit pitted Mrs. Modern, with her electric dishwasher, against Mrs. Drudge, with her old-fashioned dishwashing methods.[61] But the Carousel of Progress, with its innovative revolving audience and its audioanimatronic family, equated consumer products with science and progress in a setting of (simulated) lived environments. As its name implied, Progressland was not simply an array of appliances; it was a place—a magical realm inhabited by everyone in the audience and engineered by companies like GE.

The evolution in corporate display from products and production processes to product environments was not unique to GE. The star attraction of the Ford Motor Company's 1939 pavilion was the Ford Cycle of Production, the brainchild of Walter Teague, which employed an elaborately animated diorama in an effort to add entertainment value to the automobile manufacturing process. For the 1964 fair, Ford carried Teague's earlier rule of thumb—"the complete elimination of technical details not familiar to the public"—one step further. Emphasizing the animation while downplaying the process, Ford hired Walt Disney to design a Magic Skyway ride, which ushered visitors through time, from scenes of "dinosaurs existing in their prehistoric habitat" and cavemen "developing the rudiments of technology" to the City of Tomorrow, all while riding in new Ford convertibles.[62] Like GE's Skydome Spectacular, Ford's Magic Skyway portrayed its sponsor as part of the inevitable march of technological progress.

General Motors, the company with the most lavish exhibit budget, felt the greatest pressure to dazzle fairgoers. The highlight of the 1939 Futurama ride had been its depiction of the "highway world of 1960." By 1964 the federal interstate highway system was well under construction, although Futurama's predicted traffic-free, seven-lane motorways with hundred-mile-per-hour speed limits still seemed out of reach.[63] Where, then, could the unproblematic future be located?

Like its namesake, Futurama II promised "a look at developments that await man in the very near future—predictions solely based on fact." Once again, superhighways linked the countryside to a "modern suburban community" and a remarkably unchanged "metropolis of tomorrow." This time, however, the highways of the future ventured to "the far reaches of our planet." "As you proceed on the Futurama ride," a GM guidebook promised, "you will be impressed by its dominant concept—*mobility*. Everywhere you go, you will see man conquering new worlds." Visitors passed a succession of animated dioramas depicting "how the frontier lands of today—desert, jungle, polar regions and ocean floor—can be made livable and productive." A generation earlier, when GM's Kettering spoke of the "spirit of pioneering" and "frontiers unlimited" and when GE's Owen Young talked of "explorers setting sail for unknown lands," they were referring to their companies' researchers as the pioneers and to science as the frontier. In 1964, when GM publicists described the Ride into Tomorrow, the "pioneers" were consumers, and the "frontiers" were imaginary consumer colonies, extending markets into remote corners of the globe and even into space. The designers of the 1964 fair might obscure science and technology more than they explicated it, but in so doing, they passed the mantle of pioneering from corporate scientists and engineers to consumers.

The nature of the new frontiers that GM chose to depict, and the language its publicists used to describe them, revealed the global dimensions (as well as the inherent domesticity) of emerging corporate visions of the future. Futurama II began with scenes of "man and machines exploring the surface of the moon" and an orbiting space colony, followed by "a weather station cut deep into the Antarctic shelf." Visitors then journeyed to the ocean floor, where guests at "the beautiful Hotel Atlantis, an undersea resort," could watch an "underwater well [pumping] oil into a train of submarine tankers" and aquacopters "probing the ocean's caverns" for the "riches of the deep." Next came a "jungle deep in the tropics," where "we observe a vibrant new quality in tomorrow's jungle—the sound and look of progress. Revolutionary new earthmoving and jungle equipment are helping to construct modern highways, towns and industrial plants. Nuclear power is being used to transform this land of impenetrable vegetation into civilization—to give motion to people and usefulness to resources too long dormant." The final stop before the City of Tomorrow was the desert: "Here, too, a miracle has taken place." By channeling "the fresh fluid of life" from underground rivers and mountain dams, future consumers enjoyed automatically tended crops, "which now grow on land too long sterile."[64]

Futurama II's vision of the future carried the corporate depiction of technology and society to the next step beyond GE's Carousel of Progress. The Carousel domesticated the fable of progress into consumption habitats; the 1964 Futurama extended those protected enclaves into remote settings. GM's revised Ride into Tomorrow thus preserved the rhetoric of new frontiers by repeating the most familiar progress fable (the conquest of nature) in unfamiliar environments.

The 1964 fair may have been the last of its kind. The rhapsodic hymn to corporate futurism sounded more forced, at times even shrill, by 1964. And in the society outside of Flushing Meadows, other voices were becoming more audible. After 1964, such unqualified celebration of technological optimism would be met by a rising chorus of protest. Only a display capable of obscuring the growing disparity between corporate dreamscapes and recognized social realities could make a Futurama III possible, and world fairs lacked the resources for such lavish concoctions of hyperreality. For technological display to survive as a public relations mode in this perplexing new era, science-based corporations needed an impresario of unprecedented skill. They found their new, improved Hank Morgan in the person of Walt Disney. His importance for the 1964 fair was already comparable to that of Kettering, Teague, and Bel Geddes in 1939. For the "imagineers" at WED (Walter E. Disney), Inc., the 1964 fair provided a trial run for a permanent world's fair: EPCOT (Experimental Prototype Community of Tomorrow) at Disney World in Florida.[65]

EPCOT: Corporate Disney World

EPCOT opened in 1982, drawing heavily from the 1939 and 1964 fairs. At one end of Disney World, the world showcase pavilions simulated global travel by representing the cultures of sponsoring nations; at the other end, the corporate exhibits of Future World simulated time travel by depicting past and future uses of technology. Although many of the same sponsors appeared, EPCOT's pavilions were fewer, more permanent, more elaborate, and more automated than those at the New York fairs. Scientist demonstrators, scarce by 1964, had been eliminated; for that matter, very few people appeared as part of the exhibits at all. The use of the ride-through exhibit, which began with GM's 1939 Futurama, reached its pinnacle at EPCOT. Borrowing from the lesson of the Futurama, each pavilion devoted primary attention to the audioanimatronic view from its "ride-through vehicles." Even more than the 1964 fair's exhibitors, EPCOT's publicists stressed entertainment over education. When asked what information or messages EPCOT was designed to convey, public relations director Charlie Ridgeway replied, "I don't know that our mission in life is to deliver messages. Our mission in life is to entertain, and in so doing to provide interesting and informative and entertaining information."[66] A message, however, did await EPCOT's visitors. During Future World's initial phase of development — from its opening in 1982 until the mid-1990s, when indications of a new, revised Future World began to appear—EPCOT provided a haven for the variety of corporate futurism that had dominated the 1939 and 1964 fairs.

Future World also represented the next stage in the extension of products into simulations of lived environments. At the New York fairs, science and technology became compressed into appliances, and appliances expanded into social scenarios of consumption. At EPCOT, each pavilion transformed its products into corporate worlds, inviting visitors to identify an entire category of their everyday lives with a particular sponsor. In effect, while voyaging through exhibit worlds, Future World visitors could see themselves inhabiting global corporate visions.

Kraft was primarily a food-processing company, and United Technologies was a consortium of weapons contractors best known for nuclear submarines. Their pavilions, however, were called The Land and The Living Seas, respectively. General Motors retained the scope of Futurama II by naming its EPCOT pavilion World of Motion. Other corporations claimed worlds of their own, and more: beyond AT&T's Spaceship Earth, visitors could explore General Electric's Horizons and Exxon's Universe of Energy. Other sponsors attempted to link their products generically with the magical properties of technology: Unisys celebrated its computers with an exhibit called Backstage Magic, while Kodak called its pavilion Journey into the Imagination.

In staking out its realm, each corporate exhibitor selected the appropriate theme (transportation, communication, food production) and presented its evolution over time in a stylized fable of progress. At GM's World of Motion, transportation was equated with freedom. The ride boasted complex audioanimatronic display technology, but its whimsical depiction of the past borrowed from GM's Parade of Progress of the 1930s.[67] Visitors witnessed caricatures of the history of human mobility: cavemen with sore feet, Egyptian inventors experimenting with square and triangular wheels, ancient Romans suffering a used-chariot dealer's hype, a stagecoach attacked by Indians, a train being robbed, an early automobile causing a traffic jam by frightening the other drivers' horses.

The problems of the past, however, quickly disappeared. Each car's recorded narration assured visitors that "we have engineered marvels that take us swiftly over the land and sea, through the air, and into space itself. And still bolder and better ideas are yet to come—ideas that will fulfill our age-old dream to be free: free in mind, free in spirit, free to follow the distant star of our ancestors to a brighter tomorrow." This cosmic utopia was represented by a high-rise futurist metropolis very much like the one in GM's 1939 Futurama.

None of GM's EPCOT exhibits acknowledged that old problems might have lingered. At the 1964 Fair, GM addressed issues of "urban mobility," offering "imaginative thinking on how traffic problems of our cities can be minimized"; at EPCOT's World of Motion, traffic jams do not appear after the turn of the century. Alternatives to the internal combustion engine, explored as possible new product lines in 1964, appeared in the World of Motion pavilion only as objects of ridicule. Although EPCOT opened soon after the second of two major oil shortages, fuel efficiency was not discussed.

General Motors' omission of substantive issues was symptomatic of a problem facing all of the initial exhibitors at EPCOT's Future World. The 1964 fair updated the 1939 fair while preserving much of its technological optimism; EPCOT's designers endeavored to incorporate changes in the fable of progress but had to reconcile technological advances with the social changes that had accompanied them. Between the 1964 New York fair and EPCOT's opening, America had witnessed "the sixties" (that is, the conflation of social crises that erupted between the mid-1960s and the mid-1970s). The civil rights movement, the antiwar movement during the Vietnam War, the women's movement, the environmental movement, and the antinuclear movement convinced a number of corporate and government leaders that authority itself was under fire.[68] The 1933 and 1939 fairs tried to convince their audiences that progress was inevitable and that private enterprise was the preferred agent of that progress. Even in the heart of the Depression, such expressions of faith in science and technology evoked supportive responses. A half-century later, EPCOT's designers confronted a culture in which the very engine of progress — science and technology — was under fire.

Disney World's imagineers and publicists solved this problem largely by simply omitting troublesome historical episodes. GE's Carousel of Progress now resides at Disney World's Magic Kingdom, which is adjacent to EPCOT. The 1964 Carousel captured the family at neat, twenty-year intervals: the turn of the century, the 1920s, the 1940s, and the 1960s. The Magic Kingdom's 1980s version of the Carousel kept the first three segments intact then skipped over the troublesome 1960s altogether, landing safely in the updated "present" of the 1980s.

Ironically, while corporate displays of science and technology were becoming more feminized, focusing more on consumerism than production, women were redefining their social role — a shift that the imagineers felt compelled to address. The 1964 Carousel had conformed to postwar sexual stereotypes, portraying men and women, respectively, as breadwinners and homemakers. Aware that this model might not fly in the 1980s, EPCOT's designers updated the Magic Kingdom's Carousel with lighthearted banter over women's rights: the turn-of-the-century mom mocked dad's assertion that labor-saving devices made her life easy; the 1920s daughter said that she would like to get a job, dad answered, "It's a man's world out there," and daughter replied, "It won't always be"; and the 1940s mom was shown renovating the basement and wondering why she did not receive equal pay for housework. Then the Carousel skipped over the 1960s, ushering the audience into the 1980s. It was New Year's Eve. Dad was cooking, while mom sat at the computer. But this role reversal was entirely cosmetic. Dad was clearly not accustomed to cooking—this was a special occasion—and mom was using the computer to file recipes and adjust the house lights. This token reversal of gender roles immunized the scene against any hint of conflict—or of genuine change. Whether viewers saw the women's movement as welcome or threatening, they could find reassurance in the tranquil domestic scene before them.

Not far from the Magic Kingdom was GE's other Disney World feature, the Horizons pavilion in EPCOT's Future World. In many ways, Horizons was an update of the Carousel; even the same audioanimatronic family served as guides, this time as residents of the twenty-first century. And while the Carousel of Progress dwelt in the past, Horizons focused on the future.

The first portion of the Horizons animated ride, however, was devoted to past notions of the future. Visitors glided past slapstick depictions of Jules Verne's turn-of-the-century moon capsule, of a 1930s robot chef inadvertently trashing a kitchen, and of "the future from the 50s," with UFOs, astroburgers, and hula hoops. This tour of past "futures" concluded with the observation that "people have painted some pretty fantastic views of the future." Just as the Carousel ridiculed technological pessimists among previous generations, so Horizons mocked overoptimistic (or frivolous) past notions of the future. The net effect was to immunize the future from the past by replacing the social history of technology with a vision of technological inevitability unhindered by real events in real time. Visitors thus proceeded from amusing caricatures of past notions of the future to GE's solemn present-day version of things to come. General Electric's new, improved 1980s future was dominated by the "new towns of tomorrow—desert farms and floating cities—even colonies in space." Carousel's twenty-first century family was distributed among these colonies: the daughter was an "agricultural engineer" at a desert reclamation project, where robots had replaced farmworkers; the son lived in a space colony; and the granddaughter's boyfriend was studying marine biology at one of the sea colonies. Like GM's 1964 Ride into Tomorrow, GE's Horizons portrayed the colonization of the desert, the ocean, and outer space as the inevitable next stage in human development—a theme repeated throughout Future World.

Harvesting the desert was highlighted in Kraft's pavilion, The Land. The Living Seas featured Sea Base Alpha, a simulated ocean colony where United Technologies was characterized as a disinterested pioneer in "undersea exploration." Space colonies provided the grand culmination for the show rides at AT&T's Spaceship Earth and GM's World of Motion. What was most striking about this "cosmic colonies" vision of the future was that the settlements looked very much like the future of the 1939 fair, only more barren and remote. With no new visual iconography of progress to replace the streamlined futurism of the past, EPCOT had to rely on literal depictions of the new frontiers metaphor, with consumers as the pioneers.

Even this strategy, however, had to undergo rhetorical modification. GM's 1964 Ride into Tomorrow provided the basic formula for EPCOT's vision of the future as outposts in the desert, the ocean, and outer space; but the language of conquest that seemed appropriate in 1964 fared poorly alongside the rising environmental awareness of the 1970s. Futurama II guidebooks could talk with impunity of "man conquering new worlds" and of leveling jungles with nuclear-powered lasers to tap "resources too long dormant," but the imagineers had to provide their corporate exhibitors with either a revised vision of the future or a subtler way of presenting the old vision. By and large, EPCOT chose the latter. The conquest of nature remained in place but with an "environmentalized" vocabulary that presented the technological homogenization of nature as "natural."

Although all of the Future World exhibits relied upon this nature-technology reversal to some degree, Kraft's pavilion, The Land, embodied it most fully. Kraft's feature exhibit was a raft ride called Listen to the Land. As they drifted through five vast greenhouses of agricultural exhibits, visitors could hear an accolade to ecological awareness: only by paying attention to the earth, they were told, and by forming "a partnership with the land," could humanity prosper. Surrounded by thousands of living plants, the viewer seemed far from the high-technology images that pervaded the rest of EPCOT. Here, if nowhere else in the park, nature appeared to be dominant. Gradually, however, a very different message emerged. The first of the greenhouses, Tropics House, demonstrated techniques for converting tropical rain forests into food factories, even though "the challenge of harvesting food in this environment would make it difficult for us to live here." Kraft invited visitors to adopt the premise that, as food production technologies for agribusiness improved, large-scale agricultural development should be extended to every corner of the globe.

This assumption left a number of questions unaddressed. Since very little of the United States could be considered tropical rain forest, just where were these lands? However difficult it might have been for "us" to live there, what about the people who already lived there? What was the implied relationship between "us" and "them"? And at a time when the deforestation of the Amazon was attracting global concern, who would benefit from technologies that clear tropical rain forests for agricultural production? Under the rubric of learning from nature, Kraft's Tropics House demonstrated how nature could be harnessed as a food production machine, enhanced by the technological advances of "controlled environment agriculture." Nature and technology were reversed through a simple act of appropriation: in the realm of The Land, nature *was* a technology: a technocratic vision of the planet, therefore, was only "natural."

Nor did this vision of conquest stop at rain forests. Desert House demonstrated irrigation and crop selection techniques for making the desert bloom. Aquacell showed how the sea could be harvested when fish were treated as crops. Creative House showcased "aeroponics" and the development of a "lunar soil simulant" to prepare us for farming on the moon: "On a lunar base," a Kraft promotional pamphlet observed, "people may someday dine on moon-grown salads as they watch Earth hang in the sky." As in GE's Horizons and GM's World of Motion, The Land envisioned the future through this triumvirate of colonies. And once again the rationale for their development was unexplained. (Why have lunar farms? To feed the colonists. And why were the colonists there? To tend the farms.)

How did the cosmic colonies come to embody EPCOT's vision of the future? Were they a tacit response to public uncertainty over the direction of technological progress? Although they were invariably described as bold exploratory ventures, these remote settlements could also be viewed as shelters — exotic versions of Jay Swayze's 1964 Underground House with its dial-a-view windows. The threat of nuclear war, and even nonmilitary uses of nuclear technology, were conspicuous by their absence from the social uses of technology depicted at EPCOT. Many of the exhibitors were important defense contractors; some were important in the nuclear power industry. At the 1964 fair, these aspects of their corporate identities had been proudly included. (GM's 1964 Avenue of Progress, for example, included exhibits of "navigation and guidance systems for a variety of defense missiles," "the work of General Motors to develop a compact, highly mobile atomic reactor for military use," and examples of "how GM scientists are putting the power of atoms to work for the betterment of mankind.") Although EPCOT avoided references to these technologies, the cosmic colonies coincided with the most elaborate of proposals for surviving a nuclear "exchange."[69]

The colonies also functioned as the ultimate stage in suburbanization. In 1939, a nation struggling to recover from a decade of Depression could look to corporate technology as the key to better times. By 1964, it was becoming clear that the dramatic social transformations predicted at the 1939 fair were not, in and of themselves, likely to reach everyone. To the extent that technological changes had improved the standard of living, they accentuated the distance between the haves and the have-nots in the culture of abundance. Postwar urban flight did not follow the vision of greenbelt equality depicted by the 1939 fair's Democracity. Nor did it permit the new generation of suburbanites to escape from the unpleasant side effects of technological growth. (Jay Swayze designed his Underground House not only as a fallout shelter but also as "the ultimate in privacy," providing protection from "the hazards of modern living, including pollution, pollen, noise and radioactive fallout.")[70]

By the 1980s, EPCOT's sponsors and imagineers were faced with conflicting mandates. The trend toward depicting corporate worlds led them to connect their depictions of the future to sweeping renditions of the past—a task that lent a far more interpretive, even didactic, quality to Future World than its designers wished to acknowledge. The increasing emphasis on display as entertainment meant that these panoramic views of technology and society over time should not dwell on problems. Yet since 1964, the public had become much more aware that life in advanced industrial America was not without its difficulties. New problems had accompanied abundance, and many of the old problems had not disappeared. For the most part, the 1964 fair proposed new trajectories for suburban flight, culminating in the Moon Room, where housewives could admire minimalist furnishings designed to blend with a lunar setting. If EPCOT was to retain its characterization of the sponsoring corporations as engines of progress, the imagineers had to either find a vision of the future in which social problems were addressed or stake out an escapist future beyond the reach of such problems as the Moon Room presented in 1964.

The fact that EPCOT chose the escape route is not surprising; after all, much of American politics and culture was seeking the same thing. Since the 1930s, the corporate display of science had increasingly relied on media formats that mystified scientific and technical issues, encouraging instead a passive faith in the capacity of corporate research and development to perpetuate the national vision of progress. It was a message that Americans had been happy to hear for two generations. By the 1980s, however, social and environmental challenges to that message had caused some consternation even among corporate boards of directors. Now that people understood that corporate science and technology was little more than superstition, what was to stop them from falling into an equally superstitious distrust?

To avoid a public reassessment of the roles of business and government in developing science and technology, Americans in the 1980s required a powerful new variety of faith in this technological determinism, one that could overcome mounting evidence of social hazards and imbalances. EPCOT provided a "timeless" setting for that renewal of faith. By locating the present between fragmented pasts and imaginary futures, EPCOT exempted its corporate sponsors from the social implications of technology over time. Future World's cosmic colonies were the last imaginable places where the 1939 fair's naivete toward technology could exist without visible evidence of its social repercussions; EPCOT's trivialization of the past was the only way to exempt the myth of progress from the lessons of history. What corporate America perceived as an image problem was corrected by airbrushing the image. As a step in the development of corporate display, EPCOT reflected that concern. For corporations to envision different social uses of science and technology would have required that they also envision a different role for corporations.

Like the *Connecticut Yankee*, EPCOT expressed the historic tension between technology and social change through an exercise in time travel; and while Hank Morgan served as the guide in the 1880s, Walt Disney led the way into the 1980s. But Twain's "expert" was the reading audience's peer, traveling back to a simpler, mythic time; at EPCOT, the audience was the latter-day equivalent of those sixth-century peasants invited to marvel at what the wizardry of the future held in store. Unlike Hank, corporations in twentieth-century America had once hesitated at depicting themselves as wizards. At fairs and exhibits up through the 1930s, many corporate exhibitors were torn between dazzling the audience with magic tricks and explaining their manufacturing processes. By the 1980s, what had begun as depictive strategies—the atmospheric approach proposed by Ruel Thompson—had become ideology. Dramatizations of the wonders of science acquired ever-shinier surfaces, but emptied of content, they became screens for the projection of inferred consumer desires. Ironically, as corporate display displaced images of the pioneer-scientist with those of the consumer-frontiersperson, both the implications of applied science and the genuine needs of consumer-citizens became obscured.

"At every turn in our history," the 1964 Carousel of Progress told its visitors, "there was always someone saying, 'Turn back, turn back.' But there is no turning back. Nor for us, not for our carousel." This celebratory assertion of technological determinism remained intact when the Carousel was installed in the Magic Kingdom. In the updated 1995 version, however, these words were gone. Similarly, the pavilions of Future World began to undergo revisions that toned down or silenced the sunny assertions of a futurism now deemed obsolete. How science-based corporations would view the future, how they would depict that future, and how passively their global consumers might receive these visions were not to be beheld on any ride-through exhibits.

PART IV

Closing the Circle

JAMES KLOPPENBERG'S chapter closes the circle in two
respects. He returns to issues, and to some of the same
thinkers, evoked in David Hollinger's and Dorothy Ross's
opening chapters. But he also returns to the turn of the
century and to Progressivism and pragmatism — to, in other
words, a crucial moment in the linking of science and reform.
In the writings of the great pragmatist John Dewey and those
of the contemporary philosopher Richard Rorty (whose
grandfather was a leading late nineteenth-century reformer
and thinker), Kloppenberg finds inspiration for locating
science and social action within a fluid and diverse society
in which conflict is a central reality.

Expressing concerns about the intrusion of "instrumental
reason" (of which the state and science are representatives)
into private life, Kloppenberg rejects the notion that the
choice is between the objectivity of science and the relativism
of interpretation. He makes an eloquent appeal for "the cultivation of a critical perspective alert to the contested nature
of all knowledge but nevertheless aware of the possibility
of making judgments on the basis of pragmatic tests and
democratic procedures." In such an America, science would
be demystified but not absent. It would face the same critical
rigor it brings to bear on its own subjects of inquiry. It would

have a far different role from the one Dr. Hale imagined in 1922, in which its claims were presumed to be universal, absolute, and the key to progress. It would be one voice among many in a messier, more contentious, and richer world of social choice.

CHAPTER EIGHT

Why History Matters to Political Theory

James T. Kloppenberg

FOR MUCH of the twentieth century, American philosophers and social scientists have thirsted for certainty. The natural sciences provided an attractive model for inquiry of all sorts, generating through systematic, rigorous research a reliable body of knowledge commanding respect and assent from specialists and nonspecialists alike. Not only were scientists sure of themselves and their findings, they also seemed capable of providing blueprints for social reform that promised results every bit as predictable as their laboratory experiments. Philosophers, dissatisfied with interminable squabbles about ethics and metaphysics, eagerly joined the crusade for truth organized by logical positivists and linguistic analysts. Students of society and politics, unhappy with the disorder of merely descriptive accounts, struggled to construct a temple of social science that would contain quantifiable, verifiable, and above all, value-free knowledge capable of providing reliable guidance for social engineers. Most historians who noticed such efforts at all viewed them with a mixture of bewilderment and dismay. For their part, analytic philosophers and empirical social scientists considered historians something of an embarrassment, quaint but probably harmless storytellers who could safely be ignored if they refused to go away.

Times have changed.[1] The natural sciences, whose claims to objectivity had intimidated humanists and inspired philosophers and social scientists — and whose hold on solid knowledge had seemed so secure — fell before the historicist analysis of Thomas Kuhn.[2] As the essays in this volume demonstrate, many of the schemes for social engineering hatched by enthusiasts for science led to results ranging from disappointing to disastrous, as research findings and their application to social problems were shown to be grounded in questionable assumptions and susceptible to appropriation for ideological purposes antithetical to the scientific ideal of value neutrality. Social scientists then began to admit what practitioners of *Geisteswissenschaften* such as Wilhelm Dilthey had known since the nineteenth century: because human experience is meaningful, understanding both behavior and expression requires interpreting the complex and shifting systems of symbols through which individuals encounter the world and with which they try to cope with it. Meanings and intentions change over time and across cultures; as that realization spread, hopes of finding a universal logic or a general science of social organization faded.[3]

Marching behind the banner of hermeneutics came an influential band of scholars who challenged the ideal of objectivity—Peter Winch, Clifford Geertz, Charles Taylor, Anthony Giddens, Paul Ricoeur, Michel Foucault, Jacques Derrida, Hans-Georg Gadamer, and Jürgen Habermas—this was a new litany of saints proclaiming variations on a revolutionary gospel of interpretation. They spoke a different language from those scientists, philosophers, and social scientists who sought to escape the clutter of history. Instead of timeless principles and truth, they spoke in a dizzying new language of revolutionary paradigm shifts, incommensurable forms of life, complexities of thick description, competing communities of discourse, archaeologies of knowledge, universal undecidability of texts, inescapability of prejudices, and colonization of life worlds by an omnivorous technostructure. Each of these moves opened new possibilities for fruitful interaction between social theory and history. Unfortunately, however, as Peter Novick's study of American historians illustrates, most practicing historians have remained as oblivious to these promising developments as they had been to the threats implicit in earlier programs aspiring to achieve scientific status by escaping the confines of history.[4] The reluctance of many historians to face the consequences of these diverse developments is unfortunate, because the many-pronged attack on natural scientists' claims to certainty, objectivity, and utility present important if seldom appreciated opportunities for developing an alternative to science as a model for social investigation and political argument. Especially since social scientists on both sides of the Atlantic failed to anticipate the most significant events of the last fifty years, the collapse of communism in Eastern Europe and the disintegration of the Soviet Union, the limitations of predictive social science are now almost universally acknowledged.

In this chapter, I examine recent developments in philosophy and political theory that reflect both growing dissatisfaction with science as a model and renewed interest in historical inquiry and I indicate why, paradoxically, the more prudent and limited claims of history continue to offer an attractive, if unsteady, source of knowledge. My concluding argument to this volume of essays by historians reflecting critically on the problematic relation between science and reform in twentieth-century America is that recent challenges to the possibility of objective truth do not, as is often assumed, diminish the value of history for political theory but, instead, magnify its importance.

Bridges built between history and politics have clearly altered the field of political theory. Don Herzog characterizes "foundationalist" claims as those "grounded on principles that are (1) undeniable and immune to revision and (2) located outside society and politics." While the canon in political theory has traditionally comprised texts unapologetically premised on just such principles, foundationalism these days seems increasingly unattractive. Contextualism, which emphasizes the intimate relations between ideas and their environment, and between theory and both linguistic and political practice, now exerts an apparently irresistible appeal for theorists of both liberal and communitarian persuasions. For example, two notable recent efforts to establish the primacy of freedom as a political value, Joseph Raz's *The Morality of Freedom* and Richard Flathman's *The Philosophy and Politics of Freedom,* both concede that rights can now be discussed not in the abstract but only in the context of particular social institutions. The validity of all political principles, in Raz's words, "is limited by their background. In this way, institutions shape the principles which are designed for the guidance and remolding of these institutions."[5]

Surely the most striking example of the increasing appeal, if not the pervasiveness, of contextualism in liberal theory, though, is John Rawls' revision of the arguments originally advanced in his *Theory of Justice*. Rawls admits that any theory can provide principles appropriate only "for us" rather than for all people at all times. He concedes that such principles derive from "our public political culture itself, including its main institutions and the historical traditions of their interpretation." He acknowledges that political theorists can now offer no more than "provisional fixed points," a deliberately oxymoronic phrase that removes his principles of justice from a disembodied moment of purportedly "rational" reflection and places them firmly in the context of our contemporary liberal democratic culture. Because Rawls now admits that he is writing "to help us work out what we now think," he emphasizes that our understanding of our history, our awareness of "the plurality of incommensurable conceptions of the good" characteristic of our liberal culture, will necessarily shape our ideas about justice.[6]

By thus historicizing his project, by removing it from the realm of universality and locating it in our time and our place, Rawls pulls the rug out from under the communitarian critics inspired by Hegel or Marx who challenge the abstract nature of his rights-bearing individuals and his thin theory of the good. Moreover, the new and improved constructivist Rawls provides ammunition his champions can use. Amy Gutmann and Stephen Holmes, for example, not only declare Rawls not guilty of committing metaphysics but also charge his fuzzy-minded Hegelian critics with foundationalism and an even deadlier sin of omission, the refusal to do history. Richard Rorty quite properly criticizes the "terminal wistfulness" that marks the conclusions of studies such as Michael Sandel's *Liberalism and the Limits of Justice,* Roberto Unger's *Knowledge and Politics,* and Alasdair MacIntyre's *After Virtue.* In the words of Jeffrey Stout, "The main problem with communitarian criticism of liberal society . . . is its implicitly utopian character. . . . When you unwrap the utopia, the batteries aren't included."[7]

One of the most powerful recent statements of liberal political theory, Stephen Holmes' double-barreled attack on the critics of liberalism in *The Anatomy of Antiliberalism* and his defense of his own version in *Passions and Constraint: On the Theory of Liberal Democracy,* relies explicitly on historical analysis. Holmes demonstrates that, whereas leftists criticize and libertarians celebrate liberal theory because both consider it compatible with a strong state and redistributionist or regulatory policies, such assumptions derive from historically inaccurate characteristics of liberalism.

Carefully explicating the writings of liberal theorists such as Locke, Montesquieu, Adam Smith, Madison, and Mill, Holmes demonstrates that "liberal principles are compatible with any form of state provision that fosters individual autonomy." Historically, if not in the caricatured version of their ideas presented by their contemporary critics, eighteenth- and nineteenth-century liberals appreciated the capacity of positive state action. They embraced a normative rather than a reductionist concept of individual interest, and they articulated a view of rights more nuanced than the false polarity between negative and positive liberty. Finally, Holmes shows that liberal theorists valued democracy but wanted to make sure popular government would survive. Liberals' efforts to ensure the security of individuals expanded gradually from their resistance to arbitrary state power to their endorsement of social insurance measures consistent with those provided by twentieth-century welfare states. Holmes's argument for liberalism, in other words, depends not on thought experiments akin to Rawls's original position but on the detailed analysis of historical evidence.[8]

These days few political theorists are content to leave their ideals lying idle in space. They are quickly becoming adept at contextualist and historicist maneuvers. Michael Walzer argues for a pluralistic conception of the "spheres of justice" because our competing values are "the inevitable product of historical and cultural particularism." In his most recent book, *Thick and Thin: Moral Argument at Home and Abroad*, Walzer contrasts richly textured political and ethical arguments ("thick"), which are rooted in a particular historical tradition, with more universal principles ("thin"), which aspire to appeal across different cultures. Attractive as the latter form of argument has been since the Enlightenment, such principles transcend the details, and thus also the reality, of actual human experience. Political ideals must be connected to the meanings people attach to that experience. "Social meanings are not just *there*, agreed on once and for all," Walzer sensibly concludes. "Meanings change over time as a result of internal tension and external example; thence, they are always subject to dispute."[9]

Communitarians no less frequently than liberals have been turning away from the abstract and toward the concrete, away from the general and toward the particular, away from the universal and toward the historical. William Sullivan, for example, tries, however schematically, to relate to American historical development the criticisms of liberalism advanced in the widely read *Habits of the Heart*. William Connolly, attempting to blend elements from Foucault's and Habermas's writings, emphasizes the difficulties facing any political system — including our own — that would seek to encourage rather than stifle differences of the sort usually suppressed by structures of power. Connolly traces that difficulty to continuing assaults on private life by the institutionalization of instrumental rationality, and his analysis may owe as much to a reading of twentieth-century social history as it does to the reading of Habermas. Finally, Richard Bernstein and Thomas McCarthy have grown increasingly enthusiastic about the attractiveness of Habermas's theory of communicative rationality as Habermas has adopted more self-consciously pragmatist ideas and deliberately repudiated appeals to a transcendental form of argument.[10]

There is some question, however, whether such moves are genuine conversions or merely convenient professions of faith in a currently fashionable historicism. J. Donald Moon chides Flathman for replacing the early Rawls foundationalism with an equally suspicious claim that rights are, after all, "elemental." Stout suggests that Rawls probably would not have labored so diligently on the original position if he were concerned only to validate it on thoroughly contextualist grounds. William Galston contends that, despite Walzer's many invocations of the concrete, particular, and historical, a "lurking universalism" pervades his analysis of democratic politics. Finally, Habermas's continuing reliance on the notion of an ideal speech situation suggests to many of his critics that, despite his pragmatist rhetoric, he ultimately retreats to the universalist rationalism of the Enlightenment.[11] But if the sincerity of the conversion to contextualism of some liberals and communitarians is open to question, there can be no such doubt about the commitments of two leading communitarian theorists, Charles Taylor and his former student Michael Sandel, or that of the Trojan horse of postanalytic philosophy, the liberal pragmatist Richard Rorty.

Taylor was among the leaders of the turn toward hermeneutics, as I have noted, and he has consistently denied the philosophical coherence of attempts to escape historical analysis for an ostensibly scientific or "rational" alternative. He rejects even the possibility of creating overarching political principles of justice appropriate for all cultures: "the judgment of what is just in a particular society involves combining mutually irreducible principles in a weighting that is appropriate for the particular society, given its history, economy, and degree of integration." In *Sources of Self: The Making of the Modern Identity,* his brilliant explanation of how the sense of individual selfhood developed in modern Western cultures, Taylor undertakes to provide just such a historical account of the process whereby a sense of "inwardness," an "affirmation of ordinary life," and a view of the creative potential of self-expression together yielded a modern Western sensibility.

We moderns are capable of great depth when we connect with the sources of our commitments, the goods that lie beyond the self. Unfortunately, we are deflected from our richest philosophical resources by a shallow subjectivism, an uncritical scientism, and an empty instrumentalism. Our widely shared commitments to justice and benevolence, Taylor argues, have become detached from the religious belief in the worth of human beings that undergirded those commitments from the date of their appearance in Western culture, and for that reason their hold on our politics has become increasingly tenous. "Adopting a stripped-down secular outlook," Taylor concludes, "involves stifling the response in us to some of the deepest and most powerful spiritual aspirations that humans have conceived." Rather than asserting that reason alone can provide an irrefutable argument on behalf of his ideals, Taylor rests his case on a historical account of the way that ancient Christian virtues were transformed by, but nevertheless continue to undergird, modern ideas of the self and its moral and political obligations.

Sandel's recent account of American political thought and practice, *Democracy's Discontent: America in Search of a Public Philosophy,* represents an attempt to bring his communitarian critique of Rawls, originally launched in *Liberalism and the Limits of Justice,* down from the realm of abstraction to the firmer ground of historical analysis. Sandel develops two different arguments. First, he examines the implicit assumptions operating beneath recent judicial decisions. Whether the issue is individual rights to free speech or privacy or religious expression, or decisions concerning abortion or divorce, courts in the United States since World War II have increasingly tended to bracket conflicts between substantive values. Viewing the individual self as something separate from, and prior to, its freely chosen ends (rather than constituted by those values, as Sandel, following Taylor, would have it), judges now interpret "the Constitution as a neutral framework of rights within which persons can pursue their own ends," whatever those ends may be. This "procedural republic," Sandel argues in the second part of his book, represents a departure from the eighteenth- and nineteenth-century American tradition of civic republicanism that prized character and emphasized the importance of forming citizens to follow a path of virtue rather than freeing them to follow whatever path they choose.

Sandel's approach in *Democracy's Discontent* differs markedly from that of his earlier work. Instead of concentrating on the conceptual inadequacy of Rawls' view of the individual, Sandel traces a strand of American political discourse from the eighteenth century to the present, seeking to demonstrate with historical evidence the presence of republican concerns in the writings of thinkers from Jefferson and Madison through Jackson and Lincoln to Croly and Brandeis. Although historians may find parts of Sandel's account unconvincing because of its selectivity and lack of contextualization, there can be no mistaking the importance of history for his argument: Sandel's case for a renewed civic project depends on his ability to show convincingly the embeddedness of republican ideals of virtuous citizenship in America's past.[12]

The almost simultaneous appearance of Holmes' spirited defense of liberal theory and Sandel's equally spirited attack on it, both of which rely on the historical record despite their dramatically different interpretations of the evidence it contains concerning the relative centrality and current attractiveness of liberal and republican ideas, illustrates the change in perspective among political theorists from abstract to historical analysis, a shift equally apparent in the writings of the philosopher Richard Rorty.

Rorty's historicism has such explosive force because he attacks the citadel of philosophy from within. His insistence that philosophy can never attain scientific status is troubling enough, but perhaps even more troubling is Rorty's judgment that the grail of objective knowledge will likewise continue to elude the natural and social sciences. Rorty first established his credentials with papers discussing standard topics in the analytic tradition, and in 1967 he edited *The Linguistic Turn*, which contains a variety of essays in the prevailing styles of linguistic and logical analysis. In his introduction to that volume, however, he suggests that the conflicts between J. L. Austin's "ordinary-language" philosophy and Rudolph Carnap's logical positivism are so fundamental that they cannot be resolved. When *Philosophy and the Mirror of Nature* appeared in 1979, Rorty was widely acknowledged as one of the most incisive critics of twentieth-century Anglo-American philosophy. Here, Rorty insists that problems such as mind-body dualism, the correspondence theory of truth, theories of knowledge, theories of language, and ultimately the entire conception of a systematic philosophy devoted to finding foundations for objective knowledge rest on misconceptions. Because we make up the questions we ask, Rorty argues, we also make up the answers. In the provocative titles of his two concluding chapters, Rorty urges his fellow philosophers to move "from epistemology to hermeneutics" and to begin practicing "philosophy without mirrors." "Systematic philosophers" such as Locke and Kant, Austin and Carnap, who sought a science of knowledge that would disclose objective truth, should give way to "edifying philosophers" such as James and Dewey, who would contribute to the "conversation of the West" without promising results that philosophy can never deliver. Although others such as Hilary Putnam, Nelson Goodman, and Richard Bernstein offer variations on the theme of "historicist undoing," to use Ian Hacking's phrase, Rorty's assault seems especially dramatic because it holds out no alternative solutions.

In the conclusion of *Philosophy and the Mirror of Nature,* Rorty stakes out his heretical position: "To drop the notion of the philosopher as knowing something about knowing which nobody else knows so well would be to drop the notion that his voice always has an overriding claim on the attention of the other participants in the conversation." In his introduction to *The Consequences of Pragmatism,* a collection of essays published in 1982, Rorty refuses to back down. There can be no "extra-historical Archimedean point," he writes. We must face the unsettling realization that "there is nothing deep down inside us except what we have put there ourselves, no criterion that we have not created in the course of creating a practice, no standard of rationality that is not an appeal to such a criterion, no rigorous argumentation that is not obedience to our own conventions." Science, to put the point bluntly, is only "one genre of literature," and all of our efforts to find solid, unchanging knowledge are futile attempts to escape the contingency of the ongoing conversation in which we all participate.[13]

As Rorty realized, there was little in these claims that was really new. Even the gasps from his critics inside and outside the academy only echoed the responses that had greeted the writings of Nietzsche and James. But because the twentieth-century enthusiasm for science had overshadowed the writings of that earlier generation of radical critics, and because Anglo-American philosophers in particular had marched down a road marked "Truth" only to find James and Dewey waiting there for them, Rorty's revival of pragmatism seemed revolutionary.

Despite Rorty's penetrating critique, both *Philosophy and the Mirror of Nature* and *The Consequences of Pragmatism* were, at that time, assimilable. The attack on objectivity had by then established beachheads in the sciences and the social sciences; Continental philosophers had been coping with hermeneutics and existential phenomenology for decades. Moreover, the mood of Rorty's writings is encouraging if not downright upbeat. " 'Pragmatism,' " he proclaims, is "the chief glory of our country's intellectual tradition." Its pioneers James and Dewey wrote, as such dark pessimists as Nietzsche and Heidegger did not, "in a spirit of social hope. They asked us to liberate our new civilization by giving up the notion of 'grounding' our culture, our moral lives, our politics, our religious beliefs, upon 'philosophical bases.' They asked us to give up the neurotic Cartesian quest for certainty."[14]

Against critics from all sides who assailed him with charges of relativism, Rorty responds that the entire notion of relativism becomes incoherent when we appreciate the contingent status of all knowledge. From the pragmatist's perspective, there is nothing for "truth" to be relative to except our tradition, our purposes, and our linguistic conventions. When we have come to that realization, a calm acceptance of our condition becomes possible, and we can begin once again to take up the conversation. While pragmatism cannot offer objectivity, neither does it threaten the survival of civilization as we know it. Pragmatism, Rorty writes in an essay entitled "Solidarity or Objectivity," is "a philosophy of solidarity rather than despair." There is no reason to discard our beliefs about the natural world—or to discard our social, moral, and political values—just because we realize we have made them rather than found them. Faith in science, like our other faiths, helps us get things done. Science helps us muddle through, and it will continue to help us even after we have stopped trying to "divinize" it. In his most recent writings, notably in the introduction to *Objectivity, Relativism, and Truth,* Rorty reiterates this argument. When he suggests that antirepresentationalists like himself "see no sense in which physics is more independent of our human peculiarities than astrology or literary criticism," it seems clear he has no more patience with the pretensions of critics presuming to dispense wisdom in the shopworn slogans of postmodernism than he has for natural scientists who carry on as before three decades after Kuhn's bombshell burst.[15]

In the absence of foundations, Rorty recommends in *Consequences* that we look instead to history. We must accept "our inheritance from, and our conversation with, our fellow-humans as our only source of guidance." This is our defense against the nihilism that realists fear will follow from pragmatism. "Our identification with our community—our society, our political tradition, our intellectual heritage—is heightened when we see this community as ours rather than *nature's, shaped* rather than found, one among many which men have made." Rorty continues to emphasize the crucial role of history in his provocative book *Contingency, Irony, and Solidarity,* published in 1989. If we were to surrender our aspirations to certainty, he writes, we "would regard the justification of liberal society simply as a matter of historical comparison with other attempts at social organization." Rorty advances an argument that historians should find intriguing, even though it may prove as unsettling for unrepentant objectivists as other species of poststructuralist criticism: "I have been urging in this book that we try *not* to want something which stands beyond history and institutions. The fundamental premise of the book is that a belief can still regulate action, can still be thought worth dying for, among people who are quite aware that this belief is caused by nothing deeper than contingent historical circumstances." In Rorty's "liberal utopia," the charge of relativism loses its force and the claim that there is " 'something that stands beyond history' becomes unintelligible." [16]

Many historians, especially those most troubled by the "historicist undoing" sweeping across philosophy and political theory, will perhaps be surprised to find the burden of providing our culture with guidance dumped in their laps. How is history going to meet this challenge? What can historians offer philosophers such as Rorty and political theorists such as Holmes and Sandel, who have given up the search for timeless truths and have embraced the particularity of history? Historians who suppose that history can still offer objective truths concerning human affairs have not been paying attention. There are two problems with historians' confident assumption that they can offer an account of the past "as it really happened," in Leopold von Ranke's familiar phrase. First, Ranke's meaning was almost diametrically opposed to the meaning we usually impose on his words, as Georg Iggers explains. Ranke's German contemporaries scoffed at Ranke's romanticism, because he aimed to enable the divine spirit, manifesting itself in history, to flow directly through its conduit, the historian, without any interference on the historian's part. So when Ranke refers to the past "as it really happened," he is closer to the idealism of Hegel than to the commonsense realism of most twentieth-century historians. [17]

The second problem with historians' veneration of objectivity is more fundamental than a simple problem of translation. The unsettling developments that eroded the confidence of Kuhn in science, Geertz in anthropology, Taylor in social theory, and Rorty in philosophy left many historians untouched. As Novick demonstrates in *That Noble Dream,* many historians of the left as well as the right generally refuse to question, or even to think seriously about, the status of historical knowledge. While there are of course exceptions, radicals no less often than conservatives continue to assume that historians' access to truth is unproblematical. Many historians seem to have missed the hermeneutic turn that revolutionized scholarship, from the sciences to philosophy. If even physicists and ethicists can concede that inquiry inescapably involves interpretation, historians need to realize that we, too, cannot escape the hermeneutic circle. The past we study, as well as the present we inhabit, exhibit the same characteristics of multidimensionality, meaningfulness, and opacity that prevent natural scientists and political theorists from disclosing universal principles embedded in the world or imposed by the structure of human reason. History always involves interpretation; it never reveals simple truths.

Does that mean history has nothing to offer? If all historical knowledge is subject to revision, it might seem that any hopes for history as an alternative to science are misplaced. But just as physicists who realize they are working within a particular paradigm can still productively do what Kuhn calls "normal science," so historians need not be disabled by the realization that they are engaged in a species of hermeneutics. We historians should learn to see our study of the past as postanalytic philosophers have taught us to see the present — as an effort to solve problems. The nature of the problems, the range of possible solutions, and the degree of success attained are all questions that require interpretation. They require making the effort to recover the past as it was lived as well as trying to understand how and why things have turned out as they have. Although we cannot escape our own historical context and enter another any more than an anthropologist can become a member of another culture, we must try to penetrate the evidence of behaviors and expressions to understand the experience of those who shaped our past. This historical sensibility, as I call it elsewhere, begins in admitting the uncertainty of our own experience and ends in the provisional interpretations that historians can offer their cultures.[18]

Earlier pragmatists turned to history for precisely the same reasons Rorty now offers. In the words of Wilhelm Dilthey, "The totality of human nature is only to be found in history." The individual "can only become conscious of it and enjoy it when he assembles the mind of the past in himself." If we can attain in the present only provisional understandings that are culturally shaped, an indispensable source of those understandings is a detailed and sophisticated knowledge of the past. As pragmatists and other philosophers and political theorists turn increasingly to "our tradition" as the source and the testing ground for our ideas and our ideals, historians should become alert not only to the limits of historical knowledge but also to its potential role in the construction of a pragmatic cultural self-understanding.[19]

The role of historians in such a pragmatic conception of culture is to insist on the diversity of voices, and the centrality of conflict, in the shaping of our tradition. In contrast to a familiar argument, I suggest that we emphasize the absence of consensus in America. The very origins of liberal tolerance, as Hume argued and as contemporary scholars confirm, can be traced to the religious wars of early modern Europe. The institutionalization of human rights emerged from pragmatic compromises among groups that could not agree on the nature of the good; the language of rights that grew up to contend with older languages of virtue reflects the inability to agree on substantive conceptions of the good. In America, the sometimes competing, sometimes complementary vocabularies of Christianity, republicanism, and liberalism were present throughout the eighteenth century, and the differences among them were masked but not erased under the pressure of fighting the Revolution and launching the Republic. Throughout the nineteenth and twentieth centuries as well, competing conceptions of freedom and justice have drawn strength from the continuing vitality of contradictory and incompatible religious and political traditions. Those conflicts have prevented any single vision from dominating public life in the United States, thereby making impossible the dramatic transformation that marks European polities, in which power can be consolidated around unitary conceptions of the right and the good. The diversity of American politics rather than any liberal consensus defines "our tradition."

For that reason, efforts to derive a single meaning from our past, whether Holmes' expansive liberalism or Sandel's civic republicanism, however plausible and attractive they may be when viewed in isolation, are nevertheless bound to be unpersuasive. Readers of both Holmes's and Sandel's books might well rub their eyes in wonder, puzzled that two observers examining the same history of American political discourse come to such diametrically opposite conclusions about its meaning. Those differences indicate the dangers involved in using the past for partisan purposes.[20] Both Holmes and Sandel insightfully explicate the writings of those within their preferred traditions. But Holmes ignores the role of civic republican and religious language; Sandel ignores the importance of natural rights in American politics prior to World War II. Although both aspire to Walzer's "thick" description, neither quite achieves it. For as the most successful studies of American political thought and behavior published during the last decade have demonstrated, our history has been marked by the blending of different vocabularies and traditions and has shown the shaping pressure exerted on the abstractions of political theory by political experience.[21]

Thomas Haskell perceptively examines "the curious persistence of 'rights talk' in an age of interpretation." One could just as persuasively call attention to the equally curious persistence of "community talk," since if we take seriously the evidence of historical scholarship, the intermingling of these styles of discourse has been as clear as their competition. A liberalism without any responsibilities accompanying rights, and a communitarianism without any concern for individual liberties, are projected fantasies. The trajectories of historical change, moreover, have been as unanticipated as these competing positions have been interdependent. The Revolution, ostensibly fought to secure autonomy for virtuous citizens, unleashed a passion for individual rights that eventually threatened to trample the virtue it was intended to facilitate. The welfare state, constructed to preserve and expand the effective liberty of those whose autonomy was compromised by others' exercise of their freedom, eventually culminated in new and perhaps even more deadening forms of dependency.[22] The American tradition is one of struggles waged by partisans of competing ideals who agree only to disagree about the adequacy of every provisional compromise. Endlessly changing circumstances, needs, values, and people have generated new forms of subjugation and new schemes of reform. The unintended consequences of countless individual choices have created new problems in an endless and dizzying spiral.

All the truths of American history and politics have been, as the title of a recent study by Daniel Rodgers has it, "contested truths." Given the centrality of diversity, and given the tendency of some commentators such as Alasdair MacIntyre and Robert Bellah to despair of such fragmentation, and of others such as Michael Walzer, Jeffrey Stout, and Don Herzog to celebrate it, the widespread dissatisfaction with Rorty's conception of a "postmodern bourgeois liberalism" grounded in "our tradition" (singular) is hardly surprising. Of the numerous critics who discuss this aspect of Rorty's work, Richard Bernstein offers perhaps the most pointed challenge. Bernstein charges that Rorty "tends to gloss over what appears to be the overwhelming 'fact' of contemporary life—the breakdown of moral and political consensus, and the conflicts and incompatibility among competing social practices. . . . It is never clear why Rorty, who claims that there is no consensus about competing conceptions of the good life, thinks there is any more consensus about conceptions of justice or liberal democracy."[23] Rorty's early responses to such criticism tended to focus on Cold War themes that I do not discuss here. I instead concentrate

198 / James T. Kloppenberg

on two aspects of Rorty's recent writings on politics that are important but inconsistent, his characterization of social democracy and his separation of the spheres of public and private philosophy.

Rorty argues, quite reasonably, that dogmas concerning free enterprise and nationalization are no longer helpful in political debate. He accepts Habermas's open-ended characterization of socialism as "overcoming the . . . rise to dominance of cognitive-instrumental" interests. In Rorty's words, "We have to find a definition that commits us to both greater equality and a change in moral climate, without committing us to any particular economic setup. . . . There is nothing sacred about either the free market or about central planning; the proper balance between the two is a matter of experimental tinkering." That is a definition that not only Habermas but also Rawls would accept, as Rorty points out. It seems likely that William James and John Dewey would have endorsed it as well. Contemporary social democrats, Rorty argues, "think that Dewey and Weber absorbed everything useful Marx had to teach, just as they absorbed everything useful Plato and Aristotle had to teach, and got rid of the residue." Recently, however, Rorty seems to have become more pessimistic about what we can learn from this earlier generation: "I do not think that we liberals can now imagine a future of 'human dignity, freedom and peace.' That is, we cannot tell ourselves a story about how to get from the actual present to such a future. . . . We have no analogue of the scenario which . . . our grandfathers had for changing the world of 1900." [24]

Rorty dedicates *Contingency, Irony, and Solidarity* to "six liberals": his parents and grandparents. His maternal grandfather was Walter Rauschenbusch, a champion of the social gospel who contributed most powerfully to the creation of a social democratic alternative in America. Rauschenbusch understood that the abuse of power and the tendency toward oppression do not spring from, and cannot entirely be dissolved by, structures of economic, social, and political organization. Yet he insisted that personal benevolence alone is not sufficient "to offset the unconscious alienation created by the dominant facts of life which are wedging entire classes apart." Only a far more egalitarian and cooperative economic system would make possible an American culture embodying "the principle of solidarity and fraternity in the fundamental institutions of our industrial life." [25] Rauschenbusch did not know precisely what that ideal would look like in practice, but he knew that a long struggle stood between his world and the achievement of his goals, and he devoted his efforts as a scholar and activist toward that ideal of evolutionary Christian socialism.

Dewey likewise cherished an egalitarian and democratic ideal, and throughout his extraordinarily productive career he insisted that we would reach that goal only if we found ways not only to think pragmatically but also to re-organize our culture so that it is more receptive to demands for equality and less responsive to the will of those with power. Max Weber shared many of the ideals and aspirations that motivated Rauschenbusch and Dewey and was more sympathetic to social democracy than is usually acknowledged.[26] Yet the differences between Dewey and Weber are also important, especially in this context, because their different perspectives shed light on some of the issues that divide Rorty from the contemporary thinker who, better than any other, seems to me to embody the spirit of pragmatic social democratic scholarship: Jürgen Habermas.

Rorty insists that his philosophical disagreement with Habermas "is not reflected in any political disagreement." Rorty explicitly endorses Habermas's pragmatic version of a socialist ideal. The difference between them centers on Rorty's defense of a privatized sphere in which liberal ironists pursue personal visions without any implications for public life. Philosophers such as Nietzsche, Heidegger, and Derrida are useless for politics, Rorty admits, but they provide insights of incalculable richness for the cultivation of individual aesthetic sensibilities. Rorty states this contrast bluntly: his utopia, a " 'poeticized' culture," would give up "the attempt to unite" one's private ways of dealing with one's finitude and one's sense of obligation to other human beings."[27] There is a problem here, as Habermas and Bernstein argue. In an effort to live up to the promise of the title of this essay, I want to indicate why history matters to the political theory of pragmatism.

The problem, as I think Weber perceived more clearly than Dewey, stems from the centrality of instrumental rationality—or means-ends reasoning in which the means overshadow consideration of the ends to be achieved—in twentieth-century culture. Given the tendency of this process of rationalization to sweep beneath it all other considerations, given—to use again the phrase Rorty adopts from Habermas—"the rise to dominance of cognitive-instrumental" interests and the need to resist those interests in order to preserve and advance the different sets of values that different groups in our culture cherish in addition to efficiency, it seems that the autonomy of the private sphere that Rorty quite properly cherishes is more vulnerable than he is willing to admit. Given the expanding domain of instrumental reason, the separation of public from private philosophy facilitates what Habermas calls the further "colonization of the lifeworld," the invasion of private realms of imagination by an omnivorous technical spirit unsympathetic to private purposes and intolerant of the creativity and diversity that Rorty wants to protect.[28] Dewey and Weber differed on the likelihood that democratic cultures could resist the rule of this cognitive-instrumental rationality, but they agreed on its centrality and the dangers accompanying it.

For strictly pragmatic reasons, therefore—and not because of a commitment to Enlightenment rationalism or an irresistible urge to "go transcendental" and ascend to the realm of Habermas's ideal speech situation—I am suspicious of Rorty's privatized and poeticized utopia. Viewed historically, the institutions that make possible such privacy and poetry appear to be fragile creations that emerged not only from the imaginations of creative artists but also from the struggles of political activists. The survival of these institutions likewise depends on our culture's willingness to resist the challenges to individual experimentation that cognitive-instrumental rationality will inevitably pose.

Rorty concludes the discussion of his poeticized ideal by rejecting Habermas's rationalist metanarrative, which privileges "communicative reason" and discounts the private imagination.

> I should like to replace both religious and philosophical accounts of a suprahistorical ground or an end-of-history convergence with a historical narrative about the rise of liberal institutions and customs—the institutions and customs which were designed to diminish cruelty, make possible government by the consent of the governed, and permit as much domination-free communication as possible to take place. . . . That shift from epistemology to politics, from an explanation of the relation between "reason" and reality to an explanation of how political freedom has changed our sense of what human inquiry is good for, is a shift which Dewey was willing to make but from which Habermas hangs back.[29]

Dewey was indeed willing to make that move, and Rorty is quite right that Dewey emphasized the role of art, and the instrumental value of creativity, in shaping a democratic and pragmatic culture. But Dewey did not believe that such an aesthetic sensibility could be cultivated, or should be exercised, in complete isolation from the public sphere. Dewey thought the new form of individualism that would emerge in a thoroughly democratized culture would require the participation of every citizen in the creative and cooperative shaping of public life. It was his democratic faith in the possibility of integrating private and public life that immunized Dewey from Weber's despair. Where Weber saw only bureaucratization and instrumental rationality trailing disenchantment, Dewey envisioned the lively interaction of personal imagination and political experimentation. Rorty insists that philosophy should discard the dualisms that stand in the way of pragmatism. On pragmatic grounds, that denial of dualisms should extend to his conception of public and private spheres and the cultural role philosophers should play.

One year after the publication of Dewey's *Art and Experience,* from which Rorty quotes in the conclusion of his defense of a privatized sphere for liberal ironists and strong poets, Dewey's *Liberalism and Social Action* appeared. There he wrote,

> If the early liberals had put forth their special interpretation of liberty as something subject to historic relativity they would not have frozen it into a doctrine to be applied at all times under all social circumstances. Specifically, they would have recognized that effective liberty is a function of the social conditions existing at any time. If they had done this, they would have known that as economic relations became dominantly controlling forces in setting the pattern of human relations, the necessity of liberty for individuals which they proclaimed will require social control of economic forces in the interest of the great mass of individuals. Because the liberals failed to make a distinction between purely formal and legal liberty and effective liberty of thought and action, the history of the last one hundred years is the history of non-fulfillment of their predictions.[30]

If we value "effective liberty of thought and action," as Dewey did, and if we value "domination-free communication," as Rorty does, we should realize how difficult it has been historically to make progress toward these ideals. We should further realize, as Rauschenbusch did when his dissent was silenced during World War I, as Weber did when he saw how easily the nationalism he cherished could be transformed into a militarism he detested, and as Dewey did when he stood to the left of America's "vital center," effective liberty and domination-free communication remain goals to be achieved rather than achievements to be celebrated. This knowledge does not require stepping into a transhistorical metanarrative, it requires only the pragmatic historical understanding that reaching our goals requires more than exercises of imagination. As Stephen Toulmin argues, the resurgence of pragmatism does not mark the end of philosophy but only the return to an older and richer tradition of practical philosophy that traces its roots to Aristotle rather than Plato. Although our historicism has removed from the pragmatic perspective any confidence in a single teleology, the touchstones of pragmatism should continue to be *phronesis* and *praxis.*[31]

The forces that have slowly gathered momentum during recent decades and that have culminated in antifoundationalism, hermeneutics, contextualism, and pragmatism may have their origins in the gradual broadening of perspectives that began, paradoxically enough, at the same time that rationalism was being enthroned by the champions of Enlightenment. As Haskell argues, the shift in causal attribution that emerged over the course of the eighteenth century was among the cultural developments that eventuated over time in the abolitionist crusade. As individuals came to see themselves and their cultures as enmeshed in webs of interdependency stretching beyond the narrow boundaries of personal experience, they began to draw connections between themselves and suffering strangers. That realization, Haskell suggests, may have flowed in unexpected ways from the rise of a market economy. Thus through a process so rich in irony that it has obscured a crucial relation between the market for goods and the end of a market for persons, the emergence of capitalism not only created new forms of domination to replace those of feudalism, it also contributed indirectly to the sensibility that eventually put an end to slavery. Not merely a new sense of responsibility, however, but agitation, persuasion, and ultimately war made the difference.[32]

If we want to "create a more expansive sense of solidarity than we presently have," as Rorty urges, if we want to expand the realm of those to whom we do not want to be cruel, the study of our tradition indicates that we should not be content with altering the contours of our imaginations. I am not suggesting that we collapse the private sphere into the public, or vice versa. The distinction between the two, the significance of which Rorty exaggerates and Sandel underestimates, is fundamental to American culture. Nor am I suggesting, as some readers might infer, that history can provide detailed road maps leading us from the problems we now face into some utopia designed this time by historians. To the contrary, I want to insist on the manifold difficulties imposed by the recognition of diversity in our tradition and the inescapability of interpretation as the means to understanding it. Both of these features of our current situation complicate enormously any effort to replace science with history as a source of blueprints for social reform, and it would be a mistake to conclude that a historical sensibility can succeed where a scientific sensibility failed to provide adequate guidance.

One of the notable features of the American tradition, as the Bill of Rights illustrates and as David Hollinger argues persuasively in his essay in this volume, has been this culture's persistent mistrust of all attempts to provide authoritative and binding statements of principle, whether drawn from religion, natural law, science, or history. We cherish the chance to amend our Constitution. But neither does that realization reduce all knowledge claims to the status of opinion. We must work to keep the tenacity of our inclination toward the either/or of objectivity and relativism from blinding us to the possibility of escaping that dichotomy through the cultivation of a critical perspective alert to the contested nature of all knowledge but nevertheless aware of the possibility of making judgments on the basis of pragmatic tests and democratic procedures. Between the discredited claims to certainty, on the one hand, and the paralyzing nihilism that reduces all knowledge to willfulness, on the other, lies the fertile ground that earlier pragmatists discovered. In our enthusiasm to cultivate it, however, we should not value the privacy of strong poets above the activity of social reformers. We should resist the impulse to separate their roles rigidly, just as we should resist the other forms of dualism that historicism has taught us to mistrust.

Pragmatists in our past have been content with neither an isolated self nor a self that seeks all its satisfactions in science, art, or politics. They have sought to unite independent thinking with effective and undogmatic social action. That insurgent spirit has been the most important contribution of pragmatism to shaping American culture, and if I am reading Rorty's recent writings right, he still believes it is the most precious part of our tradition. As he writes in *The Consequences of Pragmatism*, "one can say something useful about truth" only in "the vocabulary of practise rather than theory, of action rather than contemplation."[33] I think Rauschenbusch, Dewey, and Habermas would agree, and I think the historical study of our tradition, and of the complicated interplay between the aspiration to scientific certainty and the impulse to reform, confirms that judgment.

Notes

Introduction / Ronald G. Walters / Uncertainty, Science, and Reform in Twentieth-Century America

1. George Ellery Hale, *A National Focus of Science and Research*, Reprint and Circular 39 (1922; Washington, D.C.: National Research Council, 1923).

2. Ibid., 522.

3. The words are from Eugenie C. Scott and appear on the back jacket of Paul R. Gross and Norman Levitt, *Higher Superstition: The Academic Left and Its Quarrels with Science* (Baltimore: Johns Hopkins University Press, 1994).

4. Steve Fuller, "Straightening out the Scientific Image," *Isis* 84 (1993): 542. The four books reviewed were Gernot Böhme, *Coping with Science* (Boulder, Colo.: Westview, 1992); Stephen Cole, *Making Science: Between Nature and Society* (Cambridge: Harvard University Press, 1992); Mary Midgley, *Science as Salvation: A Modern Myth and Its Meaning* (London: Routledge, 1992); and Milton A. Rothman, *The Science Gap: Dispelling the Myths and Understanding the Reality of Science* (Buffalo, N.Y.: Prometheus, 1992).

5. Fuller, "Scientific Image," 542.

6. Gross and Levitt, *Higher Superstition*.

7. Thomas S. Kuhn, *The Structure of Scientific Revolutions* (Chicago: University of Chicago Press, 1962).

8. I discuss these points in Ronald G. Walters, *American Reformers: 1815–1860* (New York: Hill and Wang, 1978), xii–xiv, 9–15.

9. This episode is recounted in Jonathan Lurie Zimmerman, " 'The Queen of the Lobby': Mary H. Hunt, Scientific Temperance, and the Dilemma of Democratic Education in America" (Ph.D. diss., Johns Hopkins University, 1993).

10. Alexis de Tocqueville, *Democracy in America*, ed. Phillips Bradley (New York: Knopf, 1945), 36.

11. There is an especially concise and helpful diagram and summary of research and development institutions in David Dickson, *The New Politics of Science* (Chicago: University of Chicago, 1988), 22–23. See also the classic by A. Hunter Dupree, *Science in the Federal Government: A History of Policies and Activities* (1957; Baltimore: Johns Hopkins University Press, 1986).

12. Dickson, *New Politics of Science,* 4. See also Olivier Zunz, "Producers, Brokers, and Users of Knowledge: The Institutional Matrix," in *Modernist Impulses in the Human Sciences, 1870–1930,* ed. Dorothy Ross (Baltimore: Johns Hopkins University Press, 1994).

13. Dickson, *New Politics of Science,* 324.

14. For a recent use of the term, see Amy Erdman Farrell, "A Social Experiment in Publishing: *Ms. Magazine,* 1972–1989," *Human Relations* 47 (1994):707–30.

15. Vannevar Bush, *Science: The Endless Frontier* (Washington, D.C.: Government Printing Office, 1945); J. Robert Oppenheimer, "Talk to Undergraduates," in *Frontiers in Science: A Survey,* ed. Edward Hutchings Jr. (New York: Basic Books, 1958), 341; Frederick Seitz, *On the Frontier: My Life in Science* (Woodbury, N.Y.: American Institute of Physics, 1994).

16. Gross and Levitt, *Higher Superstition.*

17. To trace some of these same developments in the writing of history, see Peter Novick, *That Noble Dream: "The Objectivity Question" and the American Historical Profession* (Cambridge: Cambridge University Press, 1988); Dorothy Ross, "Grand Narrative in American Historical Writing: From Romance to Uncertainty," *American Historical Review* 100 (1995): 651–77.

18. Dickson, *New Politics of Science.*

19. These episodes are recounted in Daniel J. Kevles, *In the Name of Eugenics: Genetics and the Uses of Human Heredity* (Berkeley: University of California Press, 1985), 106–7; James H. Jones, *Bad Blood: The Tuskegee Syphilis Experiment,* rev. ed. (New York: Free Press, 1993).

20. John C. Burnham, *How Superstition Won and Science Lost: Popularizing Science and Health in the United States* (New Brunswick, N.J.: Rutgers University Press, 1987), is an interesting commentary on this process as well as an important guide to understanding the relationship between science and a consumer economy.

1 / David A. Hollinger / How Wide the Circle of the "We"?

A version of this chapter was presented to a joint session of the annual meetings of the American Historical Association and the History of Science Society, Washington, D.C., December 29, 1992. For critical suggestions offered on that occasion and in response to earlier versions presented at a number of universities, I wish to thank the following colleagues: Thomas Bender, Lorraine J. Daston, Nicholas Dirks, Philip Gleason, Gerald Graff, Don Herzog, Frederick Hoxie, Martin Jay, David Joravsky, Evelyn Fox Keller, BettyAnn Kevles, Thomas W. Laqueur, Robert C. Post, Adolph Reed Jr., Richard Rorty, Richard Candida Smith, Werner Sollors, Ann Laura Stoler, and Ronald G. Walters. Several of these will wish I had followed their advice more fully.

1. Alfred C. Kinsey et al., *Sexual Behavior in the Human Male* (Philadelphia, 1948); and Alfred C. Kinsey et al., *Sexual Behavior in the Human Female* (Philadelphia, 1953).

2. See, for example, Lionel Trilling's review in the April 1948 issue of *Partisan Review*, reprinted in Lionel Trilling, *The Liberal Imagination: Essays on Literature and Society* (New York, 1950), esp. 222. Kinsey himself acknowledged that he and his staff had studied only North Americans, and of certain social categories, but he retained his sweeping titles and his zoological persona.

3. Wendell L. Willkie, *One World* (New York, 1943).

4. Edward Steichen, *The Family of Man* (New York, 1955). This enduring classic was brought out in a thirtieth-anniversary edition in 1985.

5. The trend toward what Harold R. Isaacs calls retribalization, within and beyond American society, began to be noted in the early 1970s. Isaacs was one of the first observers to comment systematically on this phenomenon in global perspective; see his *Idols of the Tribe: Group Identity and Political Change*, 2d ed. (Cambridge, Mass., 1989). At almost the same moment, a resolute defender of the older universalism produced an ambitious, popular book in relation to a thirteen-part program series for public television: Jacob Bronowski's vigorously and explicitly species-centered *The Ascent of Man* (Boston, 1973). Soon, the book and the series were all but forgotten. The unselfconscious masculinity of Bronowski's sense of the species and its history renders *Ascent of Man* an especially poignant example of a mode of thought placed sharply on the defensive immediately after Bronowski wrote it.

6. Alasdair MacIntyre, *After Virtue: A Study in Moral Theory* (Notre Dame, Ind., 1981); Robert N. Bellah et al., *Habits of the Heart: Individualism and Commitment in American Life* (Berkeley, Calif., 1985); Richard Rorty, *Contingency, Irony, and Solidarity* (New York, 1989).

7. Thomas S. Kuhn, *The Structure of Scientific Revolutions* (Chicago: University of Chicago Press, 1962).

8. Richard Rorty, "Solidarity or Objectivity," in *Post-Analytic Philosophy*, ed. John Rajchman and Cornel West (New York, 1985), 3–19.

9. In using the word *historicism* to denote this recognition of historicity, I follow a modern convention adopted by historians and most social scientists and humanistic scholars. Some philosophers and literary scholars still resist this usage and hold, instead, to the older, German idealist sense of the term as conveying a belief in an absolute meaning to history, reflecting an absolute will; see, for instance, Jonathan Ree, "The Vanity of Historicism," *New Literary History* 20 (1991): 961–83, who treats historicity of the sort I invoke as the precise opposite of historicism.

10. An unusually elegant statement of the historicity of science and the possibilities for reforming it is Paul Feyerabend, "Realism and the Historicity of Knowledge," *Journal of Philosophy* 86 (1989): 393–406. See esp. 404–6, where Feyerabend describes scientists as "sculptors of reality," capable even of peopling the world once again with gods, depending on the character of the energies invested in scientific inquiry.

11. Kuhn, *Structure of Scientific Revolutions*. I emphasize the role of this book in con-

solidating a tradition of historicist thinking in David Hollinger, "T. S. Kuhn's Theory of Science and Its Implications for History," *AHR* 78 (1973): 370–93; and in David Hollinger, "Free Enterprise and Free Inquiry: The Emergence of Laissez-Faire Communitarianism in the Ideology of Science in the United States," *New Literary History* 21 (1990): 897–919.

12. Clifford Geertz, *Local Knowledge: Further Essays in Interpretive Anthropology* (New York, 1983), 16.

13. Eric R. Wolf, *Europe and the People without History* (Berkeley, Calif., 1982), 391.

14. Perhaps the most widely discussed single manifestation of this impulse among anthropologists is James Clifford and George E. Marcus, eds., *Writing Culture: The Poetics and Politics of Ethnography* (Berkeley, Calif., 1986).

15. John Rawls, "Justice as Fairness: Political, Not Metaphysical," *Philosophy and Public Affairs* 14 (1985): 223–52; Rawls, "The Idea of an Overlapping Consensus," *Oxford Journal of Legal Studies* 7 (1987): 1–25.

16. This phrase was contributed by Charles Taylor, "Understanding in Human Science," *Review of Metaphysics* 34 (1980): 26.

17. Among the many commentaries on the modern-postmodern divide that can serve to confirm this view of it — although none explicitly advances the species-to-ethnos formulation — are Scott Lash and Jonathan Friedman, eds., *Modernity and Identity* (Oxford, 1992); Steven Best and Douglas Kellner, *Postmodern Theory: Critical Interrogations* (New York, 1991); David Harvey, *The Condition of Postmodernity: An Enquiry into the Origins of Cultural Change* (Oxford, 1989).

18. Stanley Fish, *Is There a Text in This Class? The Authority of Interpretive Communities* (Cambridge, Mass., 1980).

19. Michael Walzer, *Spheres of Justice: A Defense of Pluralism and Equality* (New York, 1983), xiv.

20. Richard Rorty, "On Ethnocentrism," *Michigan Quarterly Review* 25 (1986): 533. See also Clifford Geertz, "The Uses of Diversity," *Michigan Quarterly Review* 25 (1986): 105–23.

21. Although he does not use this language, Zygmunt Bauman may be making much the same point in his *Legislators and Interpreters: On Modernity, Post-Modernity, and Intellectuals* (Ithaca, N.Y., 1987), esp. 5. See also S. P. Mohanty, "Us and Them: On the Philosophical Bases of Political Criticism," *Yale Journal of Criticism* 2 (1989): 1–31; Jeffrey C. Alexander, "Bringing Democracy Back In: Universalistic Solidarity and the Civil Sphere," in *Intellectuals and Politics: Social Theory in a Changing World*, ed. Charles C. Lemert (London, 1991).

22. See, for example, Bernard Williams, "Left-Wing Wittgenstein, Right-Wing Marx," *Common Knowledge* 1 (1992): 42; Susan Wolf, "Comment," in *Multiculturalism and "The Politics of Recognition": An Essay by Charle Taylor*, ed. Amy Gutmann (Princeton, N.J., 1992), 85.

23. Peter Novick, *That Noble Dream: The "Objectivity Question" and the American Historical Profession* (New York, 1988), 469–71; Carl Becker's famous essay was reprinted in a book of the same title, *Everyman His Own Historian: Essays on History and Politics* (New York, 1935). Becker addresses both personal and tribal contexts for the exercise of the historical imagination. While Novick's phrase "every group its own historian" neatly catches the trend, this phrase risks leaving the mistaken impression that exclusivist ideologies characterize the bulk of scholarly work in ethnoracial and gender studies of the previous generation. Many scholars who identify with the subject matter of these studies welcome other scholars of both genders and of all ethnoracial affiliations to join in the study of "their" group.

24. Thomas S. Kuhn, "Rhetoric and Liberation," paper prepared for conference at the University of Iowa, Mach 28, 1984, as cited by Thomas L. Haskell, "The Curious Persistence of Rights Talk in the 'Age of Interpretation,'" *Journal of American History* 74 (1987): 1011.

25. For other critical discussions of Rorty's "ethnocentrism," see Giles Gunn, "Rorty's *Norum Organum*," *Raritan* 10 (1990): 80–103, esp. 99–100; Alexander Nehamas, "A Touch of the Poet," *Raritan* 10 (1990): 104–25, esp. 113–14, 119, 123; Thomas McCarthy, "Private Irony and Public Decency: Richard Rorty's New Pragmatism," *Critical Inquiry* 16 (1990): 355–70. McCarthy's piece is the basis for a later, relevant exchange: Richard Rorty, "Truth and Freedom: A Reply to Thomas McCarthy," *Critical Inquiry* 16 (1990): 633–43; Thomas McCarthy, "Ironist Theory as Vocation: A Response to Rorty's Reply," *Critical Inquiry* 16 (1990): 644–55.

26. As late as 1985, Rorty ("Solidarity or Objectivity," 13) explained that "to be ethnocentric is to divide the human race into the people to whom one must justify one's beliefs and the others. The first group, one's ethnos — comprises those who share enough of one's beliefs to make fruitful conversation possible." Rorty soon changed his tune, slightly: to advocate ethnocentric behavior "may sound suspicious," Rorty allowed rather defensively in 1987, but only "if we identify ethnocentrism with pig-headed refusal to talk to representatives of other communities." Rorty still did not say talk with such people, or learn from them, but he was moving in that direction. Adopting a much softer definition of *enthocentrism* than the one of two years before, Rorty explained that to be "ethnocentric" is "simply to work by our own lights," which are in any event the only ones we have. For these remarks of 1987, see Richard Rorty, "Science as Solidarity," in *The Rhetoric of the Human Sciences: Language and Argument in Scholarship and Public Affairs*, ed. John S. Nelson et al. (Madison, Wis., 1987), 43. See also Richard Rorty, "Thugs and Theorists," *Political Theory* 15 (1987): 565, 575, an essay that responds to a searching critique by Richard Bernstein, "One Step Forward, Two Steps Backward: Richard Rorty on Liberal Democracy and Philosophy," *Political Theory* 15 (1987): 538–63, in which Bernstein (esp. 553–54) challenges Rorty to address more critically what he means by "we."

27. Rorty, *Contingency, Irony, and Solidarity*, 196, 198. In his most recent writings, Rorty continues the trajectory of making openness, rather than closure, the meaning of

ethnocentrism: in the introduction to *Objectivity, Relativism, and Truth* (New York, 1991), Rorty defends "an *ethnos* which prides itself on its suspicion of ethnocentrism" (2).

28. Rorty, "On Ethnocentrism," 532.

29. Richard Rorty, "Postmodernist Bourgeois Liberalism," *Journal of Philosophy* 80 (1983): 588.

30. Sarah Akhtar, "Masai Women Should Benefit from Changed Way of Life," *New York Times,* November 11, 1989. This letter calls the degradation of Masai women to the attention of those who "seem to have fallen victim to the common fallacy that all underdeveloped and indigenous cultures are superior to the attempts by governmental forces to change them."

31. See, for example, the column by Ruth Rosen, "Women's Rights Are the Same as Human Rights," *Los Angeles Times,* April 8, 1991, arguing that "we must stop trivializing sexual crimes" in other societies "by calling them customs."

32. For a helpful discussion, see Myra Jehlen, "Archimedes and the Paradox of Feminist Criticism," *Signs* 6 (1981): 575–601; and, responding to Jehlen, Mohanty, "Us and Them," 28–29. See also Chandra Mohanty, Ann Russo, and Lourdes Torres, eds., *Third World Women and the Politics of Feminism* (Bloomington, Ind., 1991).

33. Rorty, *Contingency, Irony, and Solidarity,* 191.

34. For a provocative study of the rise of humanitarianism that can be regarded as a case study in the expansion of "the circle of the we" under changing economic and technological circumstances, see Thomas L. Haskell, "Capitalism and the Origins of the Humanitarian Sensibility," pts. 1 and 2, *AHR* 90 (1985): 339–61, 547–66. See also the closely related essay, Thomas W. Laqueur, "Bodies, Details, and Humanitarian Narrative," in *The New Cultural History: Essays,* ed. Lynn Hunt (Berkeley, Calif., 1989).

35. Rorty, *Contingency, Irony, and Solidarity,* 192.

36. Geertz, "Uses of Diversity," 122.

37. The term *postethnic* is used by Werner Sollors with reference to some recent studies of American literature; see Sollors' trenchant essay, "A Critique of Pure Pluralism," in *Reconstructing American Literary History,* ed. Sacvan Bercovitch (Cambridge, Mass., 1986), 270.

38. Hence, *postethnic* might be construed as bearing to *ethnic* the same relation that *post-Marxism* bears to *Marxism* in the well-known formulation of Ernesto Laclau and Chantal Mouffe: "But if our intellectual project in this book is *post*-Marxist, it is evidently also post-*Marxist.*" See Laclau and Mouffe, *Hegemony and Socialist Strategy: Towards a Radical Democratic Politics* (London, 1985), 4.

39. Ernest Gellner, *Postmodernism, Reason, and Religion* (New York, 1992); Arthur M. Schlesinger Jr., *The Disuniting of America: Reflections on a Multicultural Society* (New York, 1992).

40. Edwin Markham, "The Outwitted," in *Poems of Edwin Markham,* ed. Charles L. Wallis (New York, 1950), 18.

41. An example of obliviousness to these distinctions is the influential account of the "we" developed by Wilfrid Sellars. Sellars assumes that all moral and epistemic "we's" are communities of consent, inhabited by individuals who choose to pursue a common inquiry. His analysis is so far removed from actual, historical communities that he describes the community relevant to his analysis as "mankind generally," even if such a community exists only in the mind of an individual and even if evidence for its existence there is simply the willingness of an individual to discuss with another such a question as "what ought to be done." See Sellars, *Science and Metaphysics: Variations on Kantian Themes* (London, 1968), 220, 225.

42. This reification of particulars is criticized in one of the most ambitious and vividly etched historical critiques of contemporary culture, Harvey, *Condition of Postmodernity,* 116–17: "Postmodernism has us accepting the reifications and partitionings, actually celebrating the activity of masking and cover-up, all the fetishisms of locality, place, or social grouping, while denying that kind of meta-theory which can grasp the political-economic processes (money flows, international divisions of labour, financial markets, and the like) that are becoming ever more universalizing in their depth, intensity, reach and power over daily life."

43. The need for attention to these features of moral communities is argued compellingly by Martha Nussbaum, "Recoiling from Reason," *New York Review of Books,* Decemmber 7, 1989, 36–41; and by Jeffrey Stout, *Ethics after Babel: The Languages of Morals and Their Discontents* (Boston, 1988).

44. Ulf Hannerz, "The World in Creolisation," *Africa* 57 (1987): 546–59. See also many of the essays in the important collection, Mike Featherstone, ed., *Global Culture: Nationalism, Globalization, and Modernity* (Newbury Park, Calif., 1990), first published as vol. 7 of *Theory, Culture, and Society* (June 1990).

45. For explorations of the problematic character of ethnoracial identity, see Michael M. J. Fischer, "Ethnicity and the Postmodern Arts of Memory," in Clifford and Marcus, *Writing Culture,* 195; Anthony Appiah, " 'But Would That Still Be Me?' Notes on Gender, 'Race,' Ethnicity, as Sources of 'Identity,' " *Journal of Philosophy* 87 (1990): 493–89; Werner Sollors, *Beyond Ethnicity: Consent and Descent in American Culture* (New York, 1986). Sollors' work has done more to redirect American discussions of ethnic identity than any other single contribution since the classic work of Nathan Glazer and Daniel Patrick Moynihan, *Beyond the Melting Pot: The Negroes, Puerto Ricans, Jews, Italians, and Irish of New York City* (Cambridge, Mass., 1963). See also the valuable collection, Werner Sollors, ed., *The Invention of Ethnicity* (New York, 1989), especially Sollors's own introduction (ix–xx).

46. Joseph Campbell, *The Hero with a Thousand Faces* (Princeton, N.J., 1949). This book is one of the great artifacts of the universalist enthusiasm of the midcentury era. See Campbell's comments on human similarity and difference in his preface, viii.

47. It is sometimes suggested that the species itself is an unstable category, like race. If recognition of differential gene frequencies in certain populations of human organisms threatens to break down homo sapiens, however, it does not follow that the concept of humankind must lose its validity as a referent for science and social theory. The question of where humanity ends and something else begins is the subject of a collection of excellent papers: James J. Sheehan and Morton Sosna, eds., *The Boundaries of Humanity: Humans, Animals, Machines* (Stanford, Calif., 1991).

48. The best brief account of the issues raised by feminist analysis of science is Evelyn Fox Keller, "Gender and Science: 1990," in Keller, *Great Ideas Today, 1990* (Chicago, 1990). A helpful discussion of the epistemological orientation of recent historiography of science is Jan Golinski, "The Theory of Practice and the Practice of Theory: Sociological Approaches in the History of Science," *Isis* 81 (1990): 492–505. See also Joseph Rouse, "The Politics of Postmodern Philosophy of Science," *Philosophy of Science* 58 (1991): 607–27, esp. 624–25.

49. For feminist discussions relevant to this aspiration, see Helen E. Longino, *Science as Social Knowledge: Values and Objectivity in Scientific Inquiry* (Princeton, N.J., 1990); Sandra Harding, "After the Neutrality Ideal: Science, Politics, and 'Strong Objectivity,'" *Social Research* 59 (1992): 567–87; Frances E. Mascia-Lees, Patricia Sharpe, and Colleen Ballerino Cohen, "The Postmodernist Turn in Anthropology: Cautions from a Feminist Perspective," *Signs* 15 (1989): 7–33. See also Lynn Hankinson Nelson, *Who Knows: From Quine to a Feminist Empiricism* (Philadelphia, 1990).

50. Walzer, *Spheres of Justice,* xiv–xv. Walzer stops short of trying to parlay these basic rights into a more extensive vision of justice "applicable everywhere." Walzer's discriminations are relevant here because he manages to justify a locally focused moral philosophy without cutting off his connections to a wider human community. His *Spheres of Justice* is animated by a belief that "our own" society is capable of becoming "a society of equals." Hence, Walzer offers not another diatribe against universalism but a relatively down-to-earth, historically self-conscious effort to bring out of his own society the egalitarian potential with which that society is endowed by its ethnos.

51. Rorty often complains that Habermas too often becomes "transcendental," but Richard Bernstein and Martin Jay insist that Habermas's project, especially as recently formulated, is compatible with the historicist perspective Rorty has done so much to advance. See Richard J. Bernstein, *Beyond Objectivism and Relativism: Science, Hermeneutics, and Praxis* (Philadelphia, 1983), esp. 183; Martin Jay, review of Jürgen Habermas, *The Philosophical Discourse of Modernity,* in *History and Theory* 28 (1989): 94–112. It is indeed Habermas's universalism that Rorty continues to point to as the greatest point of disagreement between them; see Rorty's clarification of his relation to Habermas in *Contingency, Irony, and Solidarity,* 61–69, esp. 67–68. The character of Habermas's "universalism" is considered carefully in a book that has much influenced my own understanding of Habermas's project: Stephen K. White, *The Recent Work of Jürgen Habermas: Reason, Justice, and Modernity* (Cambridge, 1988).

52. Stout, *Ethics after Babel*, 258–59.

53. Michael Walzer, "What Does It Mean to Be an 'American'?" *Social Research* 57 (1990): 614.

54. See John Higham, "Hanging Together: Divergent Unities in American History," *Journal of American History* 61 (1974): 10–19; Philip Gleason, "American Identity and Americanization," in *Harvard Encyclopedia of American Ethnic Groups*, ed. Stephan Thernstrom (Cambridge, Mass., 1980). Important clarifications concerning this ideological tradition are offered by Rogers M. Smith, "The 'American Creed' and American Identity: The Limits of Liberal Citizenship in the United States," *Western Political Quarterly* 41 (1988): 225–51.

55. The vehemence of particularist initiatives in post-1989 Europe has generated in some European intellectuals a neouniversalist reaction that extends even to the celebration—"American style"—of hybridization. "I'm a passionate supporter of all that can insure that tomorrow people on this continent will be hybrids," says the French philosopher Bernard-Henri Levy, for whom *Europe*, rather than *France* or *humanity*, serves as the salient "we." Levy describes his ideal Europe as a "machine to break, complicate, reduce, and put into perspective all identities and national groups" but laments that such a Europe is generally rejected in favor of more narrowly particularist identities. "The great modern and murderous delirium in Europe is the folly of ethnic identity." Bernard-Henri Levy, quoted in "An Alarum for the New Europe Is Sounded from a Paris Salon," *New York Times*, December 13, 1992.

56. This point is not to be confused with the more extravagant claim that multicultural America is "isomorphic with the world," wisely disputed in the name of the world's prodigious diversity by George Yudíce, "We Are *Not* the World," *Social Text*, nos. 31–32 (1992): 202–16. Nor do I intend here to slight, in an "American exceptionalist" fashion, the fact that some nations other than the United States—Brazil is a convenient example—perform this mediating role.

57. Herman Melville, *White-Jacket; or, The World in a Man-of-War* (1850; New York, 1966), 158.

58. A refreshing exception to this rule of avoidance is Mitchell Cohen, "Rooted Cosmopolitanism," *Dissent* (Fall 1992): 478–83, which defends a perspective on American nationality similar to the one I sketch here.

59. The value of locating primary solidarity in citizenship within a democratic nation-state is an implication of much recent work by philosophers, historians, and political economists. The philosopher Thomas Nagel's *Equality and Partiality* (New York, 1991) cautiously argues (178) that the inherently dangerous but indispensable instinct for "solidarity" is better acted on in relation to such a nation-state than in relation to "racial, linguistic, or religious identification," on the one hand, or "the world," on the other. Historian Roman Szporluk's *Communism and Nationalism: Karl Marx versus Friedrich List* (New York, 1988) concludes (240) that nationalism can be consistent with an affirmation "as fundamental values" of "the rights of the individual and humanity's community of

fate." Political economist Robert B. Reich's *Work of Nations: Preparing Ourselves for 21st-Century Capitalism* (New York, 1991) provides a careful justification for a nation-centered answer to the question—generated by his analysis of the world capitalist economy—that forms the title of its final chapter: "Who Is 'Us'?." Reich's perspective can be instructively contrasted to the capitalism-conquers-all vision of postnationality advanced by a champion of the international business elite Kenichi Ohmae, in *The Borderless World: Power and Strategy in the Interlinked Economy* (London, 1990).

60. For a helpful discussion of what role these social distinctions should and should not play in the constitutional order of the United States, see Robert C. Post, "Cultural Heterogeneity and the Law: Pornography, Blasphemy, and the First Amendment," *California Law Review* 76 (1988): 320. Although Post addresses a highly specific issue in contemporary constitutional doctrine—the appropriateness of restriction of pornography in the context of the First Amendment—he develops along the way a lucid, general account of pluralist, assimilationist, and individualist principles in the American constitutional and political tradition. One of the many virtues of Post's analysis is his awareness of the potentially different implications of pluralism for groups defined by gender, as opposed to those defined by race and ethnicity or by voluntary religious affiliation. For an argument in a direction different from Post's, see Iris Marion Young, *Justice and the Politics of Difference* (Princeton, N.J., 1990), which posits group differences as the basis for a reformed civic culture. Young acknowledges (260) that group difference "is the source of some of the most violent conflict and repression in the world today," but she tries to develop a political theory responsive to this danger as well as to the diversifying promise of strengthened groups in American life.

61. Angela Davis, "Rope," *New York Times*, May 24, 1992. See also Cohen, "Rooted Cosmopolitanism"; Bruce Robbins, "Comparative Cosmopolitanism," *Social Text*, nos. 31–32 (1992): 169–86.

62. The ability of white ethnics to choose their ethnicity has been widely noted; see, for instance, Mary C. Waters, *Ethnic Options: Choosing Identities in America* (Berkeley, Calif., 1990).

63. In David Hollinger, "Postethnic America," *Contention* 2 (1992): 79–96, I defend this ideal in the context of the American multiculturalist debate and express my pessimism that this goal can be achieved in the foreseeable future. I try also to distinguish between universalism, multiculturalism, cultural pluralism, and cosmopolitanism.

64. Fernand Braudel, *The Mediterranean and the Mediterranean World in the Age of Philip II* (New York, 1972), 1:18. This passage was first called to my attention by James J. Sheehan.

2 / Dorothy Ross / A Historian's View of American Social Science

This chapter is based on Dorothy Ross, *The Origins of American Social Science* (Cambridge: Cambridge University Press, 1991), a study of the core social sciences in the

United States—economics, sociology, and political science—from their eighteenth-century roots to 1929. Early versions of this essay were given as the Bland-Lee Lectures at Clark University, February 17 and 18, 1992; as the keynote address at Cheiron, June 19, 1992: and as an article, published under the same title, in the *Journal of the History of the Behavioral Sciences* 29 (1993): 99–112.

1. A characteristic example of the critique from the left is C. Wright Mills, *The Sociological Imagination* (New York: Oxford University Press, 1959).

2. For an excellent summary of these developments, see Joan Williams, "Critical Legal Studies: The Death of Transcendence and the Rise of the New Langdells," *New York University Law Review* 62 (1987): 429–96. On the linguistic turn as historicism, see Richard Rorty, "Dewey between Hegel and Darwin," in *Modernist Impulses in the Human Sciences,* ed. Dorothy Ross (Baltimore: Johns Hopkins University Press, 1994), 56.

3. The claim that the social sciences originated in a historical question, namely the character and future of modern society, has generally been made for sociology and its classic nineteenth-century texts: Philip Abrams, "The Sense of the Past and the Origins of Sociology," *Past and Present,* no. 55 (1972): 18–19. That the claim holds good for the exemplary eighteenth-century texts of social science can be seen in Duncan Forbes, " 'Scientific' Whiggism: Adam Smith and John Millar," *Cambridge Journal* 7 (1954): 643–70; Keith Michael Baker, *Condorcet* (Chicago: University of Chicago Press, 1975), chap. 6; Isaiah Berlin, *Vico and Herder* (New York: Viking, 1976); Friedrich Meinecke, *Historism* (London: Routledge and Kegan Paul, 1972), chap. 3.

4. On historicism, see Meinecke, *Historism;* Hayden V. White, "On History and Historicisms," in Carlo Antoni, *From History to Sociology: The Transition in German Historical Thinking* (Detroit: Wayne State University Press, 1959); Calvin G. Rand, "Two Meanings of Historicism in the Writings of Dilthey, Troeltsch, and Meinecke," *Journal of the History of Ideas* 25 (1964): 503–18. An interesting recent study that emphasizes the historicizing effect of the Reformation is Anthony Kemp, *The Estrangement of the Past: A Study in the Origins of Modern Historical Consciousness* (New York: Oxford University Press, 1991).

5. Ross, *Origins,* 5–9; Forbes, " 'Scientific' Whiggism"; Don Herzog, *Without Foundations* (Ithaca: Cornell University Press, 1985).

6. Ross, *Origins,* chap. 1. On science as an effort to stabilize historicist time, see also Kemp, *Estrangement of the Past,* 160–62.

7. Karl Popper, *The Poverty of Historicism* (Boston: Beacon, 1957), 161.

8. Robert E. Park and Ernest W. Burgess, *Introduction to the Science of Sociology* (Chicago: University of Chicago Press, 1921), 8–12. See also Ross, *Origins,* 364–66.

9. Robert E. Park, "The City: Suggestions for the Investigation of Human Behavior in the City Environment," *American Journal of Sociology* 20 (1915): 577–79; Park and Burgess, *Introduction,* 509–11. See also Ross, *Origins,* 357–64.

10. Max Weber, *Roscher and Knies: The Logical Problems of Historical Economics* (New York: Free Press, 1975), 217 n., 54–65; Thomas Burger, *Max Weber's Theory of Concept Formation,* expanded ed. (Durham: Duke University Press, 1987); Lelan McLemore, "Max Weber's Defense of Historical Inquiry," *History and Theory* 23 (1984): 277–95.

11. Dorothy Ross, "Historical Consciousness in Nineteenth-Century America," *American Historical Review* 89 (1984): 909–28; Perry Miller, *Nature's Nation* (Cambridge: Harvard University Press, 1967); Ross, *Origins,* chap. 2. For a recent study of the liberal tradition in nineteenth-century American literature and Gilded Age social science that imaginatively examines American constructions of nature, see Howard Horwitz, *By the Law of Nature: Form and Value in Nineteenth-Century America* (New York: Oxford University Press, 1991).

12. Weber admittedly took a more strenuous attitude toward the distance required of university scholars than most current practitioners of poststructural theory. He drew a distinction between the values imbedded in perspective and language and overt evaluative judgment and sought to avoid the latter. See Christopher G. A. Bryant, *Positivism in Social Theory and Research* (New York: St. Martin's, 1985), chap. 3; Wilhelm Hennis, "The Pitiless 'Sobriety of Judgment': Max Weber between Carl Menger and Gustav von Schmoller — the Academic Politics of Value Freedom," *History of the Human Sciences,* 4 (1991): 27–59.

13. Ross, *Origins,* chap. 2.

14. Ibid., chap. 3.

15. Ibid., chap. 4.

16. Ibid., 115–22; John Bates Clark, *Essentials of Economic Theory* (New York: Macmillan, 1907), vii.

17. Henry Adams, *History of the United States of America,* 9 vols. (New York: Scribner's, 1890), 9:225. For a more extended discussion of Adams' historical consciousness, see Dorothy Ross, "Modernist Social Science in the Land of the New/Old," in Ross, *Modernist Impulses in the Human Sciences, 1870–1930.*

18. John Bates Clark, *The Distribution of Wealth* (New York: Macmillan, 1899), vi, 76, 409, 418–19.

19. John Bates Clark, *Social Justice without Socialism* (Boston: Houghton Mifflin, 1914), 14–15, 48–49; Ross, *Origins,* 202, 229, 242, 336, 387–88.

20. Ross, *Origins,* chap. 5.

21. E. R. A. Seligman, *The Economic Interpretation of History* (New York: Macmillan, 1902); Edward A. Ross, *Social Control* (New York: Macmillan, 1901); Frank J. Goodnow, *Politics and Administration* (New York: Macmillan, 1900).

22. Ross, *Origins,* chaps. 6–8.

23. Arthur F. Bentley, *The Process of Government* (1908; Cambridge: Harvard University Press, 1967), 206–8, 219–22, 243, 274, 301–5, 354–59.

24. Walton H. Hamilton, "The Development of Hoxie's Economics," *Journal of Political Economy* 24 (1916): 860; William I. Thomas, "The Persistence of Primary-Group Norms in Present-day Society and Their Influence in Our Educational System," in *Suggestions of Modern Science Concerning Education*, ed. Herbert S. Jennings et al. (New York: Macmillan, 1917), 107 and 179–96 passim. See also Ross, *Origins*, 303–30.

25. Ross, *Origins*, 312–19; Ross, "Modernist Social Science." On modernist historical consciousness, see Matei Calinescu, *Five Faces of Modernity: Modernism Avant-Garde Decadence Kitsch Postmodernism* (Durham: Duke University Press, 1987), pt. 1; Renato Poggioli, *The Theory of the Avant-Garde* (Cambridge: Harvard University Press, 1968), 66–67, 72–74, and chap. 10; Richard J. Quinones, *Mapping Literary Modernism: Time and Development* (Princeton: Princeton University Press, 1985), chaps. 1–2.

26. William F. Ogburn, *Social Change, with Respect to Culture and Original Nature* (New York: B. W. Huebsch, 1922); Park and Burgess, *Introduction to the Science of Sociology;* William I. Thomas and Florian Znaniecki, *The Polish Peasant in Europe and America*, 2 vols. (1918; New York: Knopf, 1927); Wesley Clair Mitchell, *Business Cycles: Memoirs of the University of California* 3 (1913). See also Ross, *Origins*, chaps. 9–10.

27. Thomas and Znaniecki, *The Polish Peasant*, 1:1, 15; Thomas, "Persistence of Primary-Group Norms," 107.

28. Bryant, *Positivism*, chap. 5.; see also Ross, *Origins*, chap. 10.

29. Dwight Waldo, *The Administrative State: A Study of the Political Theory of American Public Administration* (New York: Ronald, 1948), notes this seemingly cyclical pattern in political science.

30. Robert H. Frank, Thomas Gilovich, and Dennis T. Regan, "Does Studying Economics Inhibit Cooperation?" *Journal of Economic Perspectives* 7 (1993): 159–71.

31. Charles Lindblom has been one of the most powerful voices in this crusade. See Charles E. Lindblom and David K. Cohen, *Usable Knowledge: Social Science and Social Problem Solving* (New Haven: Yale University Press, 1979); and Charles E. Lindblom, *Inquiry and Change: The Troubled Attempt to Understand and Shape Society* (New Haven: Yale University Press, 1990).

3 / JoAnne Brown / Crime, Commerce, and Contagionism

The primary research for this chapter was begun under the auspices of the Smithsonian Institution postdoctoral fellowship program; I thank Ray Kondratas, Barbara Melosh, Michael Harris, John Fleming, Audrey Davis, and Charles McGovern for their support. Samuel Haber and David Hollinger commented on an early draft, "Crime and Contagion: The Popularization of Germ Theory in the United States, 1870–1950," which was prepared for the 1988 meeting of the Organization of American Historians, Reno, Nevada. Other readers from whose insight I have benefited are John Harley Warner, Barbara Gutmann Rosenkrantz, Gert Brieger, Harry M. Marks, Ronald G. Walters, Dale Smith, Owsei Temkin, Henry Tom, and the anonymous readers for the Johns Hopkins

University Press. I wish also to thank Howard Boyer for his early confidence in the larger book project for which this essay was the germ.

1. *The Germ*, 1923 melodrama feature film, directed by P. S. McGreeney, written by Charles Winton Warnack. The connections between coercion and hygiene have been explored in the French case most famously by Michel Foucault in *Discipline and Punish: The Birth of the Prison*, trans. Alan Sheridan (New York: Pantheon, 1977); and Michel Foucault, *The Birth of the Clinic: An Archaeology of Medical Perception*, trans. A. M. Sheridan Smith (New York: Vintage, 1973). See also Bruno Latour, *Les microbes: Guerre et paix suivi de irreductions* (Paris: Metaile, 1984); Lion Murard and Patrick Zylberman, "De l'hygiene comme introduction a la politique experimentale (1875-1925)," *Revue de Synthese*, 3d ser., no. 115 (1984): 313-14; David S. Barnes, *The Making of a Social Disease: Tuberculosis in Nineteenth-Century France* (Berkeley: University of California Press, 1995), 109.

2. Important early works on popular understandings of germ theory include Howard D. Kramer, "The Germ Theory and the Early Public Health Program in the United States," *Bulletin of the History of Medicine* 22 (1948): 233-47; Phyllis Allen Richmond, "American Attitudes toward the Germ Theory of Disease, 1860-1880," *Journal of the History of Medicine* 9 (1954): 428-54. See also Charles V. Chapin, *What Changes Has the Acceptance of Germ Theory Made in Measures for the Prevention and Treatment of Consumption?* (Providence, R.I., 1888), 12; James Cassedy, *Charles V. Chapin and the Public Health Movement* (Cambridge: Harvard University Press, 1962); Gert H. Brieger, "American Surgery and the Germ Theory of Disease," *Bulletin of the History of Medicine* 40 (1966): 135-45; Andrew McClary, "Germs Are Everywhere: The Germ Threat as Seen in Magazine Articles, 1890-1920," *Journal of American Culture* 3 (1980): 33-46.

3. Studies in this tradition include Naomi Rogers, "Germs with Legs: Flies, Disease, and the New Public Health," *Bulletin of the History of Medicine* 63 (1989): 599-617; Nancy Tomes, "The Private Side of Public Health: Sanitary Science, Domestic Hygiene, and the Germ Theory, 1870-1900," *Bulletin of the History of Medicine* 64 (1990): 509-39; older works such as Rene Dubos and Jean Dubos, *The White Plague: Tuberculosis, Man, and Society* (1952; New Brunswick: Rutgers University Press, 1987); Charles Rosenberg, *The Cholera Years: The United States in 1832, 1849, and 1866* (Chicago: University of Chicago Press, 1962).

4. This is not to argue that military and criminological tropes cannot be found elsewhere or earlier; they can, as Owsei Temkin has shown in "An Historical Analysis of the Concept of Infection," in *Studies in Intellectual History* (Baltimore: Johns Hopkins University Press, 1953). Nor did religious or moralistic notions of disease disappear along with metaphors of seed and soil. On agricultural imagery, see Vivian Nutton, "The Reception of Fracastoro's Theory of Contagion: The Seed that Fell among Thorns?" *OSIRIS*, 2d ser. 6 (1990): 196-234; Lester King, M.D., "The Germ Theory and Its Influence," *JAMA* 249 (1983): 794-98, 794.

5. See Chapin, *What Changes Has the Acceptance of Germ Theory Made*, 12; Barbara G. Rosenkrantz, "Cart before Horse: Theory, Practice, and Professional Image in Ameri-

can Public Health, 1870–1920," *Journal of the History of Medicine and Allied Sciences* 29 (1974): 55–73, 73; Rosenkrantz, introduction to Dubos and Dubos, *White Plague;* Naomi Rogers, "Dirt, Flies, and Immigrants: Explaining the Epidemiology of Poliomyelitis, 1900–1916," *Journal of the History of Medicine* 44 (1989): 486–505, 487; Rogers, "Germs with Legs," 616; Tomes, "The Private Side of Public Health."

6. I view these cultural patterns as undirected and uncoordinated yet coherent, referring often unselfconsciously to shared metaphoric vocabularies. The historian's retrospect reconstructs this coherence in ways that parallel the difficult but central role of intellectuals in Gramsci's political theory. See *Selections from the Prison Notebooks of Antonio Gramsci,* ed. and trans. Quintin Hoare and Geoffrey Nowell Smith (New York: International Publishers, 1971), 333–36; but also see his remarks on the contingent social definition of intellectuals: "One cannot speak of non-intellectuals" (9). On Mary Mallon and the problems of coercive public health policy, see Judith Walzer Leavitt's important article, " 'Typhoid Mary' Strikes Back: Bacteriological Theory and Practice in Early Twentieth-Century Public Health," *ISIS* 83 (1992): 608–29.

7. On the importance of culture and the specific role of intellectuals in mediating and reinforcing the coercive powers of the state, see *Selections from the Prison Notebooks of Antonio Gramsci;* Perry Anderson, "The Antimonies of Antonio Gramsci," *New Left Review* 100 (November 1976–January 1977): 5–78. The cultural diffusion of state powers is also, of course, essential to Michel Foucault's work.

8. Eric Monkkonen, *Police in Urban America, 1860–1920* (New York: Cambridge University Press, 1981), 2.

9. See John Higham, *Strangers in the Land: Patterns of American Nativism, 1860–1925* (1955; New Brunswick, N.J.: Rutgers University Press, 1992); Paul Boyer, *Urban Masses and Moral Order in America, 1820–1920* (Cambridge: Harvard University Press, 1978); Robert H. Wiebe, *The Search for Order, 1877–1920* (New York: Hill and Wang, 1967), chap. 4; Richard Sennett, *The Fall of Public Man* (New York: Knopf, 1976), 39; Alan M. Kraut, *Silent Travelers: Germs, Genes, and the Immigrant Menace* (New York: Basic Books, 1994).

10. Thomas J. Archdeacon, *Becoming American: An Ethnic History* (New York: Free Press, 1983), chap. 3; John D. Seelye, "The American Tramp: A Version of the Picaresque," *American Quarterly* 15 (1963): 535–53; Monkkonen, *Police in Urban America,* 88–89; T. J. Jackson Lears, *No Place of Grace: Antimodernism and the Transformation of American Culture* (New York: Pantheon, 1981). See also note 13, below.

11. See JoAnne Brown, "Professions," in *Companion to American Thought,* ed. James T. Kloppenberg and Richard Wightman Fox (Oxford: Blackwell, 1995); JoAnne Brown, *The Definition of a Profession* (Princeton: Princeton University Press, 1992); JoAnne Brown, "Professional Language: Words that Succeed," *Radical History Review* 34 (1986): 33–51; Paul Starr, *The Social Transformation of American Medicine* (New York: Basic Books, 1985); Martin Pernick, "Medical Professionalism," *Encyclopedia of Bioethics* 3 (1978): 1028–33; Martin Pernick, *A Calculus of Suffering: Pain, Professionalism, and An-*

aesthesia in Nineteenth-Century America (New York: Columbia University Press, 1985); Philip Pauly, "Summer Resort and Scientific Discipline: Woods Hole and the Structure of American Biology, 1882–1925," in *The American Development of Biology,* ed. R. Rainger, K. Benson, J. Maienschein (New Brunswick, N.J.: Rutgers University Press, 1988); Rosenkrantz, "Cart before Horse."

12. Brown, *Definition of a Profession,* 18–34.

13. Roland Marchand, *Advertising the American Dream: Making Way for Modernity, 1920–1940* (Berkeley: University of California Press, 1985), xxi. On the development of mass markets for consumer goods and the history of advertising, see Daniel Horowitz, *The Morality of Spending: Attitudes toward the Consumer Society in America, 1875–1940* (Baltimore: Johns Hopkins University Press, 1985); Daniel Pope, *The Making of Modern Advertising* (New York: Basic Books, 1983); Richard Wightman Fox and T. J. Jackson Lears, eds., *The Culture of Consumption: Critical Essays in American History, 1880–1930* (New York: Pantheon, 1983); Stephen R. Fox, *The Mirror Makers: A History of Advertising and Its Creators* (New York: Morrow, 1984); Michael Schudson, *Advertising, the Uneasy Persuasion: Its Dubious Impact on American Society* (New York: Basic Books, 1984); David Levine, "Consumer Goods and Capitalist Modernization," *Journal of Interdisciplinary History* 22 (1991): 67–77; Charles F. McGovern, "Sold American: Inventing the Consumer, 1890–1940" (Ph.D. diss., Harvard University, 1993).

14. On the religious and agricultural etymology of the term *propaganda,* see the entries under "propaganda" and "propagate" in the *Shorter Oxford English Dictionary,* usages dating from the early 1600s and evoking the parable of the sower. See, for example, J. A. Larrabee, "The Schoolhouse as a Factor in the Production of Disease," *JAMA* 11 (1888): 613–14, which also connects the dissemination of ideas to the spread of disease, ambiguously, as both its partner and its foe. See also JoAnne Brown, "Playing the Game: Tuberculosis, Medievalist Nostalgia, and the Great War," paper prepared for the meeting of the Organization of American Historians, Atlanta, April 1994.

15. This is a composite. See the numerous examples of this expository style in "Soap" file, Warshaw Collection of Business Americana, National Museum of American History, Smithsonian Institution.

16. Alfred Lief, *It Floats: The Story of Procter and Gamble* (New York: Rinehart, 1958).

17. Marchand dates this shift at 1911–14, but the Sapolio advertisements, selling social order, were produced between 1900 and 1906. See Marchand, *Advertising the American Dream,* 10; see also Susan Strasser, *Satisfaction Guaranteed: The Making of an American Mass Market* (New York: Pantheon, 1989).

18. For the nineteenth century, see Russell C. Maulitz, "Robert Koch and American Medicine," *Annals of International Medicine* 97 (1982): 761–66; King, "Germ Theory and Its Influence"; Tomes, "The Private Side of Public Health"; Rosenkrantz, "Cart before Horse"; Charles E. Rosenberg, *The Care of Strangers: The Rise of America's Hospital System* (New York: Basic Books, 1987), chap. 5. On the scientific history of germ theory, see especially Patricia Peck Gossel, "The Emergence of American Bacteriology, 1875–1900"

(Ph.D. diss., Johns Hopkins University, 1988). On ethnicity and the germ theory, see Kraut, *Silent Travelers;* Leavitt, " 'Typhoid Mary' Strikes Back"; and Rogers, "Dirt, Flies, and Immigrants."

19. Charles Rosenberg, *No Other Gods: On Science and American Social Thought* (Baltimore: Johns Hopkins University Press, 1976), 33.

20. See Richard L. Bushman and Claudia L. Bushman, "The Early History of Cleanliness in America," *Journal of American History* 74 (1988): 1213–38.

21. W. C. Wallace, quoted in Charles E. Rosenberg, "The Cause of Cholera: Aspects of Etiological Thought in Nineteenth-Century America," in *Sickness and Health in America: Readings in the History of Medicine and Public Health,* ed. Judith Walzer Leavitt and Ronald G. Numbers (Madison: University of Wisconsin Press, 1978), 260.

22. Rosenberg, *Cholera Years,* 77–78.

23. Richmond, "American Attitudes toward the Germ Theory of Disease." See also Chapin, *What Changes Has the Acceptance of Germ Theory Made,* 12; Brieger, "American Surgery and the Germ Theory of Disease," 135–45.

24. Elizabeth Yew, "Medical Inspection of Immigrants at Ellis Island, 1891–1924," *Bulletin of the New York Academy of Medicine* 56 (1980): 488–510, 490.

25. Ivory soap advertisement, *St. Nicholas,* 1885, Warshaw Collection; see also the advertising files at the Procter and Gamble Archive, Cincinnati, Ohio; and the warnings against cheap soap by A. B. Palmer, M.D., L.L.D., in the WCTU tract, *Hygiene for Young People* (New York, 1884), 146–47.

26. Bottle and package, William Radam's "Microbe Killer," Medical Sciences Collection, National Museum of American History; see also James Harvey Young, *The Toadstool Millionaires* (Princeton: Princeton University Press, 1961), 167–70; Starr, *The Social Transformation of American Medicine,* 128.

27. John Harley Warner, "Ideals of Science and Their Discontents in Late Nineteenth-Century American Medicine," *ISIS* 82 (1991): 454–78.

28. "How to Disinfect," C. T. Kingzett for the Sanitas Company, 1895, Warshaw Collection.

29. Lifebuoy soap advertisement, ca. 1885, Warshaw Collection. This confusion between "microbes" and "germs" was commonplace in the popular literature, although there was a technical distinction between the two in the biological understanding of the time.

30. "The What—The Why—The Way of Internal Baths," pamphlet, ca. 1890, Warshaw Collection; Susan B. Cayleff, *Wash and Be Healed: The Water-Cure Movement and Women's Health* (Philadelphia: Temple University Press, 1991).

31. Pratt's Germ-a-thol advertisement, ca. 1906, Warshaw Collection.

32. Gillette advertisement, c. 1906, Warshaw Collection.

33. Pro-phy-lac-tic advertisement, Warshaw Collection; Steven Schlossman, JoAnne Brown, and Michael Sedlak, *The Public School in American Dentistry* (Santa Monica: RAND, 1986);

34. On the history of racial imagery, see especially Winthrop D. Jordan, *White over Black: American Attitudes toward the Negro, 1550–1812* (New York: Norton, 1968); George M. Frederickson, *White Supremacy: A Comparative Study in American and South African History* (New York: Oxford University Press, 1981); Kraut, *Silent Travelers;* John Dower, *War without Mercy: Race and Power in the Pacific War* (New York: Pantheon, 1986).

35. On racial violence, see especially Ida B. Wells-Barnett, *On Lynchings: Southern Horrors; A Red Record; and Mob Rule in New Orleans* (1892, 1895, 1900; New York: Arno, 1969); Joel Williamson, *The Crucible of Race: Black-White Relations in the American South since Emancipation* (New York: Oxford University Press, 1984), chap. 6; Jacqueline Dowd Hall, " 'The Mind that Burns in Each Body': Women, Rape, and Racial Violence," in *Powers of Desire: The Politics of Sexuality,* ed. Ann Snitow, Christine Stansell, and Sharon Thompson (New York: Monthly Review Press, 1983).

36. 20-Mule Team Brigade pamphlet, ca. 1905, Warshaw Collection.

37. Ivory soap advertisement, ca. 1895, Warshaw Collection.

38. Trade card, Lautz Brothers, ca. 1890, Warshaw Collection.

39. *Ye Book of Spotless Town,* Warshaw Collection; Fox, *The Mirror Makers,* 25, notes that the four most heavily advertised products of the 1980s were Sapolio, Ivory Soap, Royal Baking Powder, and Douglas Shoes.

40. My understanding of Peiss's work is drawn from "Fighting for Looks: African-American Beauty Culture," a public lecture, sponsored by Women's Studies, that Peiss gave March 4, 1993, at Johns Hopkins University.

41. I discuss metaphor's reciprocity in Brown, *Definition of a Profession.*

42. Tera W. Hunter, "Household Workers in the Making: Afro-American Women in Atlanta and the New South, 1861–1920" (Ph.D. diss., Yale University, 1990), published as *Contesting the New South: The Politics and Culture of Wage Household Labor in Atlanta, 1860 to 1920* (Chapel Hill: University of North Carolina Press, 1996).

43. Kolynos toothpaste promotional packet, 1952, Warshaw Collection.

44. On the construction of enemies, see Kenneth Burke, *A Grammar of Motives* (New York: McGraw-Hill, 1945), 6–7; Murray Edelman, *Politics as Symbolic Action* (New York: Academic, 1971), 16–17; Edelman, *Constructing the Political Spectacle* (Chicago: University of Chicago Press, 1988), 66–89. On the converse situation, the construction of infection as a form of enemy invasion, see Temkin, "An Historical Analysis of the Concept of Infection."

45. See, on the importance of streetcar advertising, Arthur P. Kellogg, "Advertising a Tuberculosis Exhibition," *Journal of the Outdoor Life* 6 (1909): 12–15; Reginard P. Stindolph, "Fighting Tuberculosis with Advertising," *Journal of the Outdoor Life* 7 (1910):

27; Robin D. G. Kelley, "We Are Not What We Seem: Rethinking Black Working-Class Opposition to the Jim-Crow South," *Journal of American History* 80 (1993): 75–112.

46. The public health concerns about streetcars are reminiscent of worries about public baths. See Georges Vigarello, *Concepts of Cleanliness,* trans. Jean Birrell (New York: Cambridge University Press, 1988); Alain Corbin, *The Foul and the Fragrant: Odor and the French Social Imagination* (Cambridge: Harvard University Press, 1986).

47. Lewis W. Hunt, *The People vs TB* (Chicago: TB Institute of Chicago and Cook County, 1966), 29; "Plain-Clothes to Look After 'Spitters,'" *Courier-Journal* (Louisville, Ky.), December 23, 1909.

48. *Ye Book of Spotless Town;* Fox, *Mirror Makers,* 45–46.

49. *Ye Book of Spotless Town.*

50. Higham, *Strangers in the Land,* remains the best and most comprehensive work on nativism in this country.

51. See especially Rogers, "Germs with Legs."

52. The notion of integration propaganda is taken from Jacques Ellul, *Propaganda: The Formation of Men's Attitudes* (New York: Random House, 1965), 74–76. Marchand also adopts this concept in *Advertising the American Dream,* xviii.

53. *The Crusader of the Wisconsin Anti-Tuberculosis Association,* no. 15 (1911): 4, cited in Lawrence Flick Papers, pamphlets, vol. 114, College of Physicians of Philadelphia.

54. See, for instance, Robert J. Newton, "The Enforcement of Anti-Spitting Laws," in *Transactions of the NASPT* (Philadelphia: William Fell, 1910); "Want a Spit Warden," *Journal of the Outdoor Life* 6 (1909): 146.

55. Rogers, "Germs with Legs," fig. 3.

56. "Film 'Health Twins at Work' Reissued," *Bulletin of the National Tuberculosis Association* 14 (1928): 24.

57. On the medical inspection of schoolchildren, see Lillian D. Wald, "Medical Inspection of Public Schools," *Annals of the American Academy of Political and Social Science* 25 (1905): 297–98; Luther Halsey Gulick and Leonard P. Ayres, *Medical Inspection of Schools* (New York: Charities Publication Committee, 1908); Thomas A. Storey, ed., *Transactions of the Fourth International Congress on School Hygiene* (Buffalo: Courier Company of Buffalo, 1913); Lewis Terman and John Almack, *The Hygiene of the School Child* (New York: Houghton Mifflin, 1929); S. Josephine Baker, *Child Hygiene* (New York: Harper, 1925); John Duffy, *A History of Public Health in New York City, 1866–1966,* (New York: Russell Sage Foundation, 1968), chap. 10; Schlossman, Brown, and Sedlak, *Public School in American Dentistry;* Brown, *Definition of a Profession,* 52–54, 58–60; Kraut, *Silent Travelers,* 226–54.

58. Kraut, *Silent Travelers,* 66.

59. Ibid.

60. See Anne Emmanuelle Birn, "Snapshot Diagnosis: The Medical Inspection of Immigrants at Ellis Island, 1900–1914" (Undergraduate honors thesis, Harvard University, 1986).

61. Dr. Alfred C. Reed, "The Medical Side of Immigration," *Popular Science Monthly* 80 (1912): 383–404, 392.

62. Ibid.

63. On the reciprocity of metaphor, see Max Black, *Models and Metaphors* (Ithaca: Cornell University Press, 1962), 40; Brown, *Definition of a Profession*, 104–5.

64. Dr. Alfred C. Reed, "Immigration and the Public Health," *Popular Science Monthly* 83 (1913): 313–38, 317; Yew, "Medical Inspection of Immigrants at Ellis Island."

65. Leavitt, " 'Typhoid Mary' Strikes Back."

66. George Rosen, *From Medical Police to Social Medicine: Essays on the History of Health Care* (New York: Science History Publications, 1974), 120–54, 201–17. Isabel Hull has recently concluded that medical police schemes were never fully enacted in Germany but remained political fantasies found only in prescriptive literature; see her *Sexuality, State, and Civil Society in Germany, 1700–1815* (Ithaca: Cornell University Press, 1996), chap. 4.

67. Monkkonen, *Police in Urban America*; Barbara Gutmann Rosenkrantz, *Public Health and the State: Changing Views in Massachusetts, 1842–1936* (Cambridge: Harvard University Press, 1972), 209.

68. "The Jolly Health Cops," 1909, Medical Sciences Archive, National Museum of American History.

69. Rosenkrantz, "Cart before Horse"; James A. Tobey, *Public Health Law: A Manual of Law for Sanitarians* (Baltimore: Williams and Wilkins, 1926), chap. 3.

70. Monkkonen, *Police in Urban America*, app. A.

71. I thank Harry Marks for suggesting this point.

72. On coercion and persuasion, see Anderson, "The Antinomies of Antonio Gramsci."

73. Charles V. Chapin, cited in Rosenkrantz, "Cart before Horse," 63–64.

74. Charles V. Chapin, *The Sources and Modes of Infection* (New York: Wiley, 1910).

75. Ibid., 166–67.

76. Ibid., 136–38.

77. Leavitt, " 'Typhoid Mary' Strikes Back."

78. The conflation of "germ" and "germ plasm" dominated eugenical anti-immigration arguments. See for this position Terence V. Powderly, "The Immigrant Menace to National Health," *North American Review* 175 (1902): 53–60.

79. "Microbion Maximum," *Journal of the Outdoor Life* 3 (1906): 91.

80. Dr. J. T. Rotrock, before the Harrisburg Academy of Medicine, as cited in "Tuberculosis—The Mont Alto Camp and Its Cures," *Telegraph* (Harrisburg, Penn.), November 30, 1904, Jacobs Scrapbook, 12/1/04–10/3/05, Henry Barton Jacobs Collection, Chesney Medical Archives, Johns Hopkins Medical Institutions, Baltimore.

81. A. Cressy Morrison, "The Transmission of Disease by Money," *Popular Science Monthly* 75 (1910): 86–88; McClary, "Germs Are Everywhere."

82. Emmet F. Pearson and Wyndham Miles, "Disinfection of Mail in the United States," *Bulletin of the History of Medicine* 54 (1980): 111–24.

83. "Tuberculosis." On washerwomen as symbols of danger, see Hunter, *Contesting the New South*. I thank Hunter for generously sharing this work before publication.

84. Tobey, *Public Health Law,* introduction.

85. J. E. Schmidt, *Dictionary of Medical Slang* (Springfield, Ill.: Charles C. Thomas, 1959), 187; cartoon, "Difference between a T. B. Bird and a Jailbird," *Journal of the Outdoor Life* 17 (1920): 113.

86. Camp Chat, *Spunk* 2 (1910): 29.

87. Maurice Buckley, "A Trip to Civilization," *Spunk* 3 (1911): 13–14.

88. Black, *Models and Metaphors,* 40; Murray Edelman, *Political Language: Words that Succeed and Policies that Fail* (New York: Academic, 1977), chap. 4; Brown, *Definition of a Profession,* 3–17, 104–5.

89. S. A. Knopf, "The Tuberculosis Problem in Prisons and Reformatories," *New York Medical Journal* (1906), reprint, S. A. Knopf Papers, National Library of Medicine, Rockville, Md. See also Dubos and Dubos, *White Plague,* 249, n. 21, on Samuel Butler's 1872 tract, *Erewhon.*

90. On inequities of social policy, see especially Edelman, *Political Language.* On similar inequalities in medicine, see especially Pernick, *Calculus of Suffering;* Rosenberg, *Cholera Years,* 60; Allan M. Brandt, *No Magic Bullet: A Social History of Venereal Disease in the United States since 1880* (New York: Oxford University Press, 1987).

91. This analysis is indirectly informed by that of Francis Fox Piven, *Regulating the Poor: The Functions of Public Welfare* (New York: Pantheon, 1971).

92. See especially Kathleen Blee, *Women of the Klan* (Berkeley: University of California Press, 1991); also Evelyn Brooks Higgenbotham, "African-American Women and the Metalanguage of Race," *Signs* 17 (1992): 251–74.

93. See Stuart Galishoff, "Germs Know No Color Line: Black Health and Public Policy in Atlanta, 1900–1918," *Journal of the History of Medicine and Allied Sciences* 40 (1985); Vanessa Gamble, ed., *'Germs Know No Color Line': Blacks and American Medicine, 1900–1945* (New York: Garland, 1989).

94. See Alice Willard Solenberger, *One Thousand Homeless Men: A Study of Original Records* (New York: Sage, 1911); John D. Seelye, "The American Tramp: A Version of the Picaresque," *American Quarterly* 15 (1963): 535–53.

95. See Katherine Ott, "The Intellectual Origins and Cultural Form of Tuberculosis in the United States, 1870–1925" (Ph.D. diss., Temple University, 1990), 92–114; Esmond Long, "Weak Lungs on the Santa Fe Trail," *Bulletin of the History of Medicine* 8 (1940): 1041–54; John Baur, *The Health Seekers of Southern California, 1870–1900* (San Marino, Calif.: Huntington Library, 1959); Billy Mac Jones, *Health Seekers in the South West, 1817–1900* (Norman: Oklahoma University Press, 1967).

96. Rosenberg, *Cholera Years,* 60; Hunter, *Contesting the New South,* chap. 5.

97. On racist imagery, see Jordan, *White over Black;* on lynching, see Williamson, *Crucible of Race;* Jacqueline Dowd Hall, *Revolt against Chivalry: Jesse Daniel Ames and the Women's Crusade against Lynching* (New York: Columbia University Press, 1979).

98. Marion Hamilton Carter, "The Vampire of the South," *McClure's Magazine* 33 (1909): 631; Charles Nesbitt, "The Health Menace of the Alien Races," *World's Work* 28 (1913): 74–75.

99. Williamson, *Crucible of Race,* 214.

100. 94. See JoAnne Brown, "The Color of Contagion: Germ Theory, Tuberculosis, and the Semantics of Segregation," paper prepared for History of Medicine Section colloquium, Yale University, November 2, 1995. See also Hunter, *Contesting the New South,* chap. 5.

101. Vivian Nutton, "The Seeds of Disease: An Explanation of Contagion and Infection from the Greeks to the Renaissance," *Medical History* 27 (1983): 1–34; Nutton, "Reception of Fracastoro's Theory of Contagion"; King, "Germ Theory and Its Influence," 796; Erwin H. Ackerknecht, "Anticontagionism between 1821 and 1867," *Bulletin of the History of Medicine* 22 (1948): 562–93.

102. See "Tuberculosis Exhibitions: The Truth Brought Home to 300,000 People in Different Cities in the Year 1906," *Journal of the Outdoor Life* 4 (1907): 27; "Bill Board Fight on Tuberculosis to Be Nation-Wide. Million Posters to Show Terrors of Disease and How to Fight It. Pictures to Be Striking," *New York American,* November 14, 1909, Jacobs Collection, Welch Library, Johns Hopkins University Medical Institutions; Livingston Farrand, "National Association: Half-Mile of Publicity," *Journal of the Outdoor Life* 6 (1909); Reginald Stindolph, "Fighting Tuberculosis with Advertising," *Journal of the Outdoor Life* 7 (1910): 271–73; Jane Addams, cited in Michael Teller, *The Tuberculosis Movement: A Public Health Campaign in the Progressive Era* (Westport, Conn.: Greenwood, 1988), 62–63.

103. William H. Allen, cited in *Journal of the National Medical Association* 3 (1911): 282.

104. Rosenkrantz, "Cart before Horse"; Ott, "Intellectual Origins"; John Harley Warner, "Exploring the Inner Labyrinths of Creation: Popular Microscopy in Nineteenth-Century America," *Journal of the History of Medicine and Allied Sciences* 37 (1982): 7–33;

Margaret Warner, "Hunting the Yellow Fever Germ: The Principle and Practice of Etiological Proof in Late Nineteenth-Century America," *Bulletin of the History of Medicine* 59 (1985): 361–65.

105. Brown, *Definition of a Profession.*

106. Thomas L. Livermore, *Numbers and Losses in the Civil War* (Boston: Houghton Mifflin, 1900); *The Medical and Surgical History of the War of the Rebellion*, vol. 6 (Washington, D.C., 1870); P. M. Ashburn, *A History of the Medical Department of the United States* (Boston: Houghton Mifflin, 1929); Stewart Brooks, *Civil War Medicine* (Springfield, Ill.: Charles C. Thomas, 1966); *The Sanitary Commission of the United States Army: A Succinct Narrative of Its Works and Purposes* (1864; New York: Arno, 1972), v–vi. The importance of the Civil War and the Great War to the popular and medical development of germ theories, and the martial vocabulary that accompanied this history, is explored in JoAnne Brown, "Matters of Life and Death: Chronic Illness and Political Culture in the United States, 1865–1945," unpublished manuscript; Brown, "Tuberculosis: A Romance," paper prepared for the Berkshire Conference of Women Historians, Vassar College, Poughkeepsie, N.Y., 1993; Brown, "Playing the Game."

107. *Sanitary Commission of the United States*, 11.

108. Rogers, "Germs with Legs."

109. Advertisement, Red Cross Absorbent Cotton, *Ladies Home Journal*, February 1917, 51. Note that by capitalizing the word *Germ* the copywriter emphasized its association with the German enemy.

110. Many physicians — most of them over fifty years of age — practicing in 1917 would have completed their medical training before 1891, when courses on bacteriology were first offered. Because many medical schools had been forced out of business after 1910, the numbers of graduates fell dramatically, and older physicians were proportionally better represented in the profession. It is likely that no small number, then, remained unconvinced of the new theories, all the while practicing antiseptic and aseptic routines. In 1917 the *American Journal of Public Health* denounced such unenlightened physicians as hygienic sinners, condemning along with the careless waitress and negligent barber "the good old doc who still believes there ain't no such thing as germs." See "Hygienic Sinners," *American Journal of Public Health* (August 1917): 720.

111. Eric J. Leed, *No Man's Land: Combat and Identity in World War I* (New York: Cambridge University Press, 1979), 19.

112. Brandt, *No Magic Bullet*, 73ff; see also David M. Kennedy, *Over Here: The First World War and American Society* (New York: Oxford University Press, 1980); Ruth Rosen, *The Lost Sisterhood: Prostitution in America, 1900–1918* (Baltimore: Johns Hopkins University Press, 1982), 33–36.

113. Naomi Rogers, *Dirt and Disease: Polio before FDR* (New Brunswick, N.J.: Rutgers University Press, 1993); Alfred W. Crosby, *America's Forgotten Pandemic: The Influenza of 1918* (Cambridge: Cambridge University Press, 1989).

114. Lysol advertisement, *Ladies Home Journal*, August 1917, 65.

115. Lysol advertisement, *Ladies Home Journal*, April 1918, 110, and in *Ladies Home Journal*, October 1917, 107.

116. Lysol advertisement, *Ladies Home Journal*, April 1917, 103.

117. Lifebuoy advertisement, 1910, Warshaw Collection.

118. Poster, "The Cold that Hangs On," Health Publicity Service, 1914, Warshaw Collection.

119. David J. Pivar, *Purity Crusade: Sexual Morality and Social Control, 1868–1900* (Westport, Conn.: Greenwood, 1973); Rosen, *The Lost Sisterhood*; Brandt, *No Magic Bullet*.

120. William Lee Howard, "How to Make Yourself Germ-Proof," *Munsey's Magazine* (1912).

121. A "rat" was a mass of saved hair used to supplement growing hair in the Gibson Girl bouffant style popular in the 1910s.

122. For usage of the term *white slavery*, and ethnic ascriptions of male criminality, see Rosen, *Lost Sisterhood*, 112–35.

123. "Jimmy Germ," *Hygeia*, September 1923, 397.

124. Reed, "Immigration and the Public Health," 325; "The Medical Side of Immigration," *Popular Science Monthly* (1912): 383–404.

125. For more thorough treatments of this subject, see Donald K. Pickens, *Eugenics and the Progressives* (Nashville: Vanderbilt University Press, 1968); Mark Haller, *Eugenics: Hereditarian Attitudes in Social Thought* (New Brunswick: Rutgers University Press, 1963); Daniel Kevles, *In the Name of Eugenics: Genetics and the Uses of Human Heredity* (Berkeley: University of California Press, 1985). A brief discussion of the medical metaphors of eugenicists is found in Brown, *Definition of a Profession*, 79–87.

126. Frances J. Hassencall, "Harry H. Laughlin, 'Expert Eugenics Agent' for the House Committee on Immigration and Naturalization, 1921–1931" (Ann Arbor, Mich.: University Microfilms, 1971); Kevles, *In the Name of Eugenics*, 103.

127. This framework is taken from Andrew Abbott, *The System of Professions: An Essay in the Expert Division of Labor* (Chicago: University of Chicago Press, 1988); but see also my arguments on medicalization in Brown, *Definition of a Profession*.

128. Untitled editorial celebrating "Wisconsin's new eugenic marriage law," *Journal of Education*, August 14, 1913.

129. On borrowed language, see Brown, *Definition of a Profession*, esp. 3–4, 23–24, 97, 104–5, 110–21, 123, 138.

130. For an example of this persistence, see Lysol advertisement, *Ladies Home Journal*, March 1952, 52. For an early feminist critique, see Rachel Lynn Palmer and Sarah Koslow Greenberg, *Facts and Frauds in Woman's Hygiene* (New York: Vanguard, 1936), 3.

131. Listerine advertisement, *Ladies Home Journal*, April 1917, 96. See also "Lysol," *Ladies Home Journal*, March 1928, 131, and *Ladies Home Journal*, April 1928, 74.

132. Johnson and Johnson antiseptic soaps advertisement, *Ladies Home Journal*, Oct. 1917, 104. The word *safe* is also a reassurance about product quality, meaning "harmless."

133. Lysol advertisement, *Ladies Home Journal*, ca. 1927, Warshaw Collection.

134. Mum advertisement, *Ladies Home Journal*, May 1937, 69.

135. Mum advertisement, *Ladies Home Journal*, June 1937, 48, and *Ladies Home Journal*, November 1937, 48.

136. Corbin, *The Foul and the Fragrant*, 43–48.

137. Frederick Hollick, *The Marriage Guide, or Natural History of Generation: A Private Instructor for Married Persons and Those about to Marry, Both Male and Female*, 200th ed.(New York: The American News Company, 1860), 163.

138. Lysol advertisement, *Ladies Home Journal*, March 1952, 52.

139. Charles V. Chapin, *How to Avoid Infection* (Cambridge: Harvard University Press, 1917); C. E-A. Winslow, *The Evolution and Significance of the Modern Public Health Campaign* (New Haven: Yale University Press, 1923), 57–58.

140. By emphasizing this secularization, I do not imply that the moral and religious meanings of cleanliness were subsumed by legalistic ones; rather, these metaphorical equations allowed the older meanings to remain attached to newer ones through a chain of metaphor. The older meanings were always held in reserve insofar as the material links in the chain could be recalled. On the political metaphors that pervaded the advertising profession during the first half of the twentieth century, see McGovern, "Sold American."

141. Hazel McGee Bowman, "A Syncopated Health Trial," *Hygeia* (October 1930): 944–46. See also discussion of the American Social Hygiene Association's cartoon film, *Health Twins at Work*, in "Film 'Health Twins at Work' Reissued," 24.

142. Kleenex advertisement, *Hygeia* (March 1933): 281. See also Leavitt, " 'Typhoid Mary' Strikes Back."

143. Bernard Berhend, "What Your Hands Tell," *Hygeia* (June 1930): 524–25.

144. Lysol advertisement, *Ladies Home Journal*, September 1937, 70. For earlier images of houseflies as criminals, see Rogers, "Germs with Legs."

4 / Allan M. Brandt / "Just Say No"

A slightly different version of this essay appears in Allan M. Brandt and Paul Rozin, eds. *Morality and Health* (New York and London: Routledge, 1997). The present essay appears with the permission of Routledge.

1. On the demographic impact of epidemics, see, for example, William H. McNeill, *Plagues and Peoples* (Garden City, N.Y.: Anchor, 1976).

2. See J. F. Fries, "Aging, Natural Death, and the Compression of Morbidity," *New*

England Journal of Medicine 303 (1980): 130–35; also, Kathy Helzlsouer and Leon Gordis, "Risks to Health in America," unpublished manuscript.

3. Mary Douglas and Aaron Wildavsky, *Risk and Culture* (Berkeley: University of California Press, 1982).

4. On the nature of epidemic disease, see, for example, Charles E. Rosenberg, *The Cholera Years: The United States in 1832, 1849, and 1866* (1962; Chicago: University of Chicago Press, 1987); John Duffy, *Epidemics in Colonial America* (Baton Rouge: Louisiana State University Press, 1953).

5. On the debate concerning the contagious nature of epidemic diseases during the nineteenth century, see Erwin H. Ackerknecht, "Anticontagionism between 1821 and 1867," *Bulletin of the History of Medicine* 22 (1948): 562–93.

6. Alfred Crosby, *Epidemic and Peace, 1918* (Westport, Conn: Greenwood, 1976), 207; also, June E. Osborn, ed., *History, Science, and Politics: Influenza in America, 1918–1976* (New York: Prodist, 1977).

7. Walsh McDermott, "Demography, Culture, and Economics and the Evolutionary Stages of Medicine," in *Human Ecology and Public Health*, ed. Edwin D. Kilbourne and William G. Smillie (New York: Macmillan, 1969).

8. Charles E. Rosenberg, "Disease in History: Frames and Framers," *Milbank Quarterly* 67 (1989): 1–15.

9. George Rosen, *A History of Public Health* (New York: MD Publications, 1958); also, Charles-Edward Amory Winslow, *The Conquest of Epidemic Disease: A Chapter in the History of Ideas* (1943; Madison: University of Wisconsin Press, 1980); Harry F. Dowling, *Fighting Infection: Conquests of the Twentieth Century* (Cambridge: Harvard University Press, 1977).

10. Rene Dubos, *The Mirage of Health: Utopias, Progress, and Biological Change* (New York: Harper and Row, 1959). See also Elliot G. Mishler, Lorna R. Amara Sigham, Stuart T. Hauser, Ramsay Liem, Samuel D. Osherson, and Nancy E. Waxler, *Social Contexts of Health, Illness, and Patient Care* (New York: Cambridge University Press, 1981).

11. Thomas McKeown, *The Role of Medicine: Dream, Mirage, or Nemesis?* (Princeton: Princeton University Press, 1979); Thomas McKeown, *The Origins of Human Disease* (Oxford: Blackwell, 1988).

12. McKeown, *Role of Medicine;* also, McDermott, "Demography, Culture, and Economics."

13. See especially McKeown, *Role of Medicine.*

14. This perspective is emphasized by Susan Sontag in her widely influential *Illness as Metaphor* (New York: Viking, 1976). Sontag claims that, as science comes to adequately account for a disease, the disease loses its metaphors or moral significance. This somewhat positivist perspective fails to recognize that new and different metaphors often replace the older ones.

15. On the nature of stigmatized disease, see, for example, Mark Edward Lender

and James Kirby Martin, *Drinking in America: A History* (New York: Free Press, 1982); Allan M. Brandt, *No Magic Bullet: A Social History of Venereal Disease in the United States since 1880* (1985; New York: Oxford University Press, 1987); and Gerald Grob, *Mental Illness and American Society, 1875–1940* (Princeton: Princeton University Press). The classic sociological treatise on the phenomena is Erving Goffman, *Stigma: Notes on the Management of Spoiled Identity* (Englewood Cliffs, N.J.: Prentice-Hall, 1963).

16. On the history of Alcoholics Anonymous, see Nan Robertson, *Getting Better: Inside Alcoholics Anonymous* (New York: Morrow, 1988); also, Jack H. Mendelson and Nancy K. Mello, *Alcohol: Use and Abuse in America* (Boston: Little, Brown, 1985); E. M. Jellinek, *The Disease Concept of Alcoholism* (New Brunswick, N.J.: Hillhouse., 1960).

17. Renee C. Fox, "The Medicalization and Demedicalization of American Society," *Daedalus* 107 (1977): 9–22.

18. Ronald Bayer, *Homosexuality and American Psychiatry: The Politics of Diagnosis* (1981; Princeton: Princeton University Press, 1987).

19. Irving R. Kaufman, "The Insanity Plea on Trial," *New York Times Magazine*, August 8, 1982, 16–20; and Janet Tighe, "A Question of Responsibility" (Ph.D. diss., University of Pennsylvania, 1987).

20. John C. Burnham, "American Medicine's Golden Age: What Happened to It?" *Science* 215 (1982): 1474–79.

21. See Brandt, *No Magic Bullet;* Bernard Dixon, *Beyond the Magic Bullet: The Real Story of Medicine* (New York: Harper and Row, 1978).

22. See, for example, Paul De Kruif, *The Microbe Hunters* (New York: 1926). De Kruif provided the scientific background for Sinclair Lewis's *Arrowsmith*.

23. Barbara Gutmann Rosenkrantz, "The Search for Professional Order in Nineteenth-Century Medicine," in *Sickness and Health in America: Readings in the History of Medicine and Public Health*, ed. Judith Walzer Leavitt and Ronald L. Numbers (Madison: University of Wisconsin Press, 1985). On the rise of the modern hospital, see Charles E. Rosenberg, *The Care of Strangers: The Rise of America's Hospital System* (New York: Basic Books, 1987).

24. Paul Starr, *The Social Transformation of American Medicine: The Rise of a Sovereign Profession and the Making of a Vast Industry* (New York: Basic Books, 1982); see also Allan M. Brandt, "The Ways and Means of American Medicine," *Hastings Center Report* 13 (1983): 41–43.

25. Lisa Berkman and Lester Breslow, *Health and Ways of Living: The Alameda County Study* (New York: Oxford University Press, 1983). A. S. Evans, "Causation and Disease: The Henle-Koch Postulates Revisited," *Yale Journal of Biology and Medicine* 49 (1976): 175–95; also Mervyn Susser, *Causal Thinking in the Health Sciences: Concepts and Strategies in Epidemiology* (New York: Oxford University Press, 1973); and Dubos, *The Mirage of Health*.

26. See, for example, Martin S. Pernick, "Politics, Parties, and Pestilence: Epidemic,

Yellow Fever, and the Rise of the First Party System," in Walzer Leavitt and Numbers, *Sickness and Health in America.*

27. Robert L. Crawford, "Cultural Influences on Prevention and the Emergence of a New Health Consciousness," in *Taking Care: Understanding and Encouraging Self-Protective Behavior,* ed. Neil D. Weinstein (New York: Cambridge University Press, 1987), 98.

28. James T. Patterson, *The Dread Disease: Cancer and Modern American Culture* (Cambridge: Harvard University Press, 1987).

29. On this movement of social epidemiology, see M. W. Flinn, *The Sanitary Condition of the Labouring Populations of Great Britain* (1832; Edinburgh: University of Edinburgh Press, 1975); William Coleman, *Death Is a Social Disease* (Madison: University of Wisconsin Press, 1984); Barbara G. Rosenkrantz, *Public Health and the State: Changing Views in Massachusetts, 1842–1936* (Cambridge: Harvard University Press, 1972); and Rosen, *History of Public Health.*

30. See, for example, James H. Cassedy, *Charles V. Chapin and the Public Health Movement* (Cambridge: Harvard University Press, 1962); also Barbara G. Rosenkrantz, "Cart before Horse: Theory, Practice, and Professional Image in American Public Health, 1870–1920," *Journal of the History of Medicine and Allied Sciences* 29 (1974): 55–73.

31. Abraham M. Lilienfeld and David E. Lilienfeld, *Foundations of Epidemiology* (New York: Oxford University Press, 1980).

32. See Raymond Pearl, "Tobacco Smoking and Longevity," *Science* 87 (1938): 216–17. Annual deaths from lung cancer rose from almost none in the late nineteenth century to 4,000 in 1935, 11,000 in 1945, and 36,000 in 1960. By the mid-1980s lung cancer was the most prevalent of all cancers, accounting for more than 150,000 deaths annually.

33. The classic epidemiological studies of smoking and cancer are Richard Doll and A. Bradford Hill, "A Study of the Aetiology of Carcinoma of the Lung," *British Medical Journal* 2 (1952): 1271–86; Doll and Hill, "The Mortality of Doctors in Relation to Their Smoking Habits: A Preliminary Report," *British Medical Journal* 1 (1954): 1451–55; Doll and Hill, "Lung Cancer and Other Causes of Death in Relation to Smoking: A Second Report on the Mortality of British Doctors," *British Medical Journal* 2 (1956): 1071–81. E. Cuyler Hammond and Daniel Horn, "Smoking and Death Rates: Report on Forty-four Months of Followup on 187,783 Men. I. Total Mortality," *Journal of the American Medical Association* 166 (1958): 1159–72.

34. U.S. Department of Health, Education, and Welfare, Public Health Service, *Smoking and Health Report of the Advisors Committee to the Surgeon General,* PHS Publication 1103 (Washington, D.C: Government Printing Office, 1964); also, Susser, *Causal Thinking in the Health Sciences.*

35. A. Lee Fritschler, *Smoking and Politics: Policy Making and the Federal Bureaucracy* (New York: Appleton-Century-Crofts, 1969).

36. Allan M. Brandt, "The Cigarette Risk and American Culture," *Daedalus* 119 (1990):155–76.

37. William B. Kannel, *The Framingham Study: An Epidemiological Investigation of Cardiovascular Disease* (Washington, D.C.: National Institutes of Health, 1968); Thomas Royle Dawber, *The Framingham Study: The Epidemiology of Atherosclerotic Disease* (Cambridge: Harvard University Press, 1980).

38. N. B. Belloc and Lester Breslow, "The Relation of Physical Health Status and Health Practices," *Preventive Medicine* 1 (1972): 409–21; Lisa Berkman and Lester Breslow, *Health and Ways of Living: The Alameda County Study* (New York: Oxford University Press, 1983).

39. John H. Knowles, "The Responsibility of the Individual," *Daedalus* 106 (1977): 57–80.

40. U.S. Department of Health and Human Services, *Healthy People: The Surgeon General's Report on Health Promotion and Disease Prevention* (Washington, D.C.: Government Printing Office, 1979)

41. Stephen Nissenbaum, *Sex, Diet, and Debility in Jacksonian America: Sylvester Graham and Health Reform* (Chicago: Dorsey, 1980).

42. James C. Wharton, *Crusaders for Fitness: The History of American Health Reformers* (Princeton: Princeton University Press, 1982). Also, Harvey Green, *Fit for America: Health, Fitness, Sport, and American Society* (Baltimore: Johns Hopkins University Press, 1986).

43. Muriel Gillick, "Health Promotion, Jogging, and the Pursuit of the Moral Life," *Journal of Health Politics, Policy, and Law* 9 (1984): 369–88.

44. Arthur J. Barsky, *Worried Sick: Our Troubled Quest for Wellness* (Boston: Little, Brown, 1988); Jane Fonda, *Jane Fonda's Workout Book* (New York: Simon and Schuster, 1981).

45. On the medicalization of American culture, see, for example, Fox, "Medicalization and Demedicalization"; Edward Shorter, *Bedside Manners: The Troubled History of Doctors and Patients* (New York: Simon and Schuster, 1985); and Irving K. Zola, "Medicine as an Institution of Social Control," *Sociological Review* 20 (1972): 487–504.

46. George Sheehan, *Running and Being* (New York: Simon and Schuster, 1979).

47. Gallup "Physical Fitness," *Gallup Report* 226 (1984): 9–12; "Attitudes about Alcohol," *Public Opinion* 12 (1989): 30–34. Also, Barsky, *Worried Sick*, 80–105.

48. See, for example, Joan Jacobs Brumberg, *Fasting Girls: The Emergence of Anorexia Nervosa as a Modern Disease* (Cambridge: Harvard University Press, 1988); also, Bruce Haley, *The Healthy Body and Victorian Culture* (Cambridge: Harvard University Press, 1978). On the general issue of self-improvement, see Donald Meyer, *The Positive Thinkers: A Study of the American Quest for Health, Wealth, and Personal Power, from Mary Baker*

Eddy to Norman Vincent Peale (Garden City, N.Y.: Doubleday, 1965); Christopher Lasch, *The Culture of Narcissism: American Life in an Age of Diminishing Expectations* (New York: Warner Books, 1979); and Lasch, *The Minimal Self: Psychic Survival in Troubled Times* (New York: Norton, 1984).

49. On the general theme of social dislocation and responses during the late nineteenth and early twentieth centuries, see especially Robert Wiebe, *The Search for Order* (New York: Hill and Wang, 1967); and Samuel P. Hays, *The Response to Industrialization* (Chicago: University of Chicago Press, 1957). On the relation of social concerns and health, see Wharton, *Crusaders for Fitness;* Martha H. Verbrugge, *Able-Bodied Womanhood: Personal Health and Social Change in Nineteenth-Century Boston* (New York: Oxford University Press, 1988).

50. Rashi Fein, "Social and Economic Attitudes Shaping American Health Policy," *Milbank Quarterly* 58 (1980): 349–85.

51. David Mechanic, "Promoting Health," paper prepared for Rockefeller Conference on the Health Transition, American Academy of Arts and Sciences, June 1989.

52. See David P. Willis, ed., *Health Policies and Black Americans* (New Brunswick, N.J.: Transaction, 1989).

53. See Brandt, *No Magic Bullet;* Elizabeth Fee and Daniel M. Fox, eds., *AIDS: The Burdens of History* (Berkeley: University of California Press, 1988).

54. Rashi Fein, *Medical Care, Medical Costs: The Search for a Health Insurance Policy* (Cambridge: Harvard University Press, 1987); Roy Lubove, "The Right to Health Care: Ethical Imperatives versus Enforceable Claims," in *Compulsory Health Insurance: The Continuing American Debate,* ed. Ronald L. Numbers (Westport, Conn.: Greenwood, 1982).

55. Leon Kass, "Regarding the End of Medicine and the Pursuit of Health," *Public Interest* 40 (1975): 11–42.

5 / Regina Morantz-Sanchez / Female Science and Medical Reform

1. For similar approaches to the physician's role, see John Eyler, *Victorian Social Medicine: The Ideas and Methods of William Farr* (Baltimore: Johns Hopkins University Press, 1979), esp. 1–12, 123–58; Gerald Grob, *Edward Jarvis and the Medical World of Nineteenth-Century America* (Knoxville: University of Tennessee Press, 1978), 3–8; Lloyd B. Stevenson, "Science down the Drain: On the Hostility of Certain Sanitarians to Animal Experimentation, Bacteriology, and Immunology," *Bulletin of the History of Medicine* 129 (1955): 1–26.

2. Charles Rosenberg, "The Therapeutic Revolution: Medicine, Meaning, and Social Change in Nineteenth-Century America," in *The Therapeutic Revolution: Essays on the*

Social History of American Medicines, ed. C. Rosenberg and M. Vogel (Philadelphia: University of Pennsylvania Press, 1979).

3. Ibid. This discussion is indebted as well to John Harley Warner, *The Therapeutic Perspective: Medical Practice, Knowledge, and Identity in America, 1820–1885* (Cambridge: Harvard University Press, 1986).

4. Henry Hartshorne, Valedictory Address, Women's Medical College of Pennsylvania, Philadelphia, 1872, 6–7.

5. Warner, *Therapeutic Perspective,* 258; Russell Maulitz, " 'Physician versus Bacteriologist': The Ideology of Science in Clinical Medicine," in Rosenberg and Vogel, *Therapeutic Revolution.*

6. Elizabeth Blackwell, "Why Hygienic Congresses Fail" (1891), reprinted in *Essays in Medical Sociology,* 2 vols. (1902; New York: Arno, 1972), 2:43.

7. Ibid., 68.

8. Warner, *Therapeutic Perspective,* 249. Also see Lucy Candib, "Ways of Knowing in Family Medicine: Contributions from a Feminist Perspective," *Family Medicine* 20 (1988): 133–36.

9. Cited by J. Goodfield, "Humanity in Science: A Perspective and a Plea," *Science* 197, quoted in George L. Engel, "How Much Longer Must Medicine's Science Be Bound by a 17th-Century World View?" in *The Task of Medicine: Dialogue at Wickenburg,* ed. Kerr L. White (Menlo Park, Calif.: Henry J. Kaiser Foundation, 1988).

10. Elizabeth Blackwell, "Scientific Method in Biology" (1898), *Essays in Medical Sociology* 2:126–30.

11. See Eric J. Cassel, "The Nature of Suffering and the Goals of Medicine," *New England Journal of Medicine* 306 (1982): 645.

12. See Regina Morantz, "Feminism, Professionalism, and Germs: A Study of the Thought of Elizabeth Blackwell and Mary Putnam Jacobi," *American Quarterly* 34 (1982): 459–78; and Regina Morantz-Sanchez, *Sympathy and Science: Women Physicians in American Medicine* (New York: Oxford University Press, 1985).

13. George L, Engel, "The Need for New Medical Model: A Challenge for Biomedicine," *Science* 196 (1977): 129, 134; Engel, "How Much Longer?"

14. H. R. Holman, "The 'Excellence' Deception in Medicine," *Hospital Practice* 11 (1976): 11, 18, 21; see also Sue Fisher, *In the Patient's Best Interest: Women and the Politics of Medical Decisions* (New Brunswick: Rutgers University Press, 1986).

15. Mary Ann Elston studies women's role in the antivivisection movement in England in detail in "Gender, Medicine, and Morality: A Study of the Antivivisection Movement, 1870–1904" (Master's thesis, University of Essex, 1984). See also Richard D. French, *Antivivisection and Medical Science in Victorian Society* (Princeton: Princeton University Press, 1975); James Turner, *Reckoning with the Beast; Animals, Pain, and*

Humanity in the Victorian Mind (Baltimore: Johns Hopkins University Press, 1980); Stevenson, "Science down the Drain." Blackwell shared with many of her contemporaries a fundamentally religious belief in divine order and the potential in each individual of an intuitive moral sense. I do not treat the religious sources of her beliefs in this chapter, but they were significant.

16. Elizabeth Blackwell, "The Influence of Women in the Profession of Medicine" (1889), *Essays in Medical Sociology* 2:13; Blackwell, "Erroneous Method in Medical Education" (1896), *Essays in Medical Sociology* 2:10–12.

17. Blackwell, "Scientific Method in Biology," 2:104.

18. See French, *Antivivisection and Medical Science*, 277–78: also Blackwell, Correspondence with Mrs. S. Wolcott Browne, especially Browne to Blackwell, May 28, 1904, Blackwell MSS, Box 55, Library of Congress.

19. Blackwell, "Influence of Women," 2:15; Blackwell, "Scientific Method in Biology," 2:105.

20. Blackwell, "Influence of Women," 2:14. See Warner, *Therapeutic Perspective*, 258–83, for an insightful discussion of these issues.

21. Blackwell, "Influence of Women," 2:19, 21.

22. Ruth Bleier, "Sex Differences in Research: Science or Belief?" in *Feminist Approaches to Science*, ed. Ruth Bleier (New York: Pergamon, 1986).

23. Elizabeth Blackwell, *Pioneer Work in Opening the Medical Profession to Women* (New York, 1896), 253. See also Morantz-Sanchez, *Sympathy and Science*, 47–63.

24. Blackwell, "Influence of Women," 2:9–10; Erik Erikson, *Childhood and Society* (New York: Norton, 1950), 267.

25. Blackwell, "Influence of Women," 2:12.

26. See Caroline Merchant, *The Death of Nature* (New York: Harper and Row, 1980), 164–90; also Elizabeth Fee, "Women's Nature and Scientific Objectivity," in *Woman's Nature: Rationalizations of Inequality*, ed. Marion Lowe and Ruth Hubbard (New York: Pergamon, 1983).

27. The literature on the social construction of scientific knowledge grows rapidly; I cite here only a representative sample: Thomas Kuhn, *The Structure of Scientific Revolutions* (Chicago: University of Chicago Press, 1962); Charles Webster, *The Great Instauration* (New York: Holmes and Meier, 1975); Keith Thomas, *Religion and the Decline of Magic* (New York: Scribner's, 1971); Sherry B. Ortner, "Is Female to Male as Nature Is to Culture?" in *Woman, Culture, and Society*, ed. M. Z. Rozaldo and L. Lamphere (Stanford: Stanford University Press, 1974): Evelyn Fox Keller, *Reflections on Gender and Science* (New Haven: Yale University Press, 1985); Bleier, *Feminist Approaches to Science;* Ruth Blier, *Science and Gender* (New York: Pergamon, 1984); Elizabeth Fee, "Science and the Woman Problem: Historical Perspectives," in *Sex Differences*, ed. Michael Teitelbaum (New York: Anchor, 1976); L. J. Jordanova, "Natural Facts: A Historical Perspective

on Science and Sexuality," in *Nature, Culture, and Gender,* ed. Carol MacCormack and Marilyn Strathern (Cambridge: Cambridge University Press, 1980); Sandra Harding, *The Science Question in Feminism* (Ithaca: Cornell University Press, 1986); Carol McMillan, *Women, Reason, and Nature* (Princeton: Princeton University Press, 1982); Ann Fausto-Sterling, *Myths of Gender* (New York: Basic Books, 1985).

28. Sara Ruddick, "Maternal Thinking," *Feminist Studies* 6 (1980): 342–67. See also Keller, *Reflections,* 67–127; McMillan, *Woman, Reason, and Nature;* Nancy Hartstock, "The Feminist Standpoint: Developing the Ground for a Specifically Feminist Historical Materialism," in *Discovering Reality,* ed. S. Harding and M. B. Hintikka (Dordrecht, Holland: Reidel, 1983); Carol Gilligan, *In a Different Voice* (Cambridge: Harvard University Press, 1982); M. F. Field, B. Clinchy, N. R. Goldberger, and J. M. Tarule, *Women's Ways of Knowing* (New York: Basic Books, 1986); Robert J. Lifton, "Woman as Knower: Some Psychohistorical Perspectives," in *The Woman in America,* ed. Robert J. Lifton (Boston: Beacon, 1967).

29. Keller, *Reflections,* 116–17.

30. Candib, "Ways of Knowing," 134.

31. Hilary Rose, "Beyond Masculinist Realities: A Feminist Epistemology for the Sciences," in Bleier, *Feminist Approaches to Science;* Nancy Hartstock, *Money, Sex, and Power: Toward a Feminist Historical Materialism* (New York: Longmans, 1983); Alfred Sohn-Rethel, *Intellectual and Manual Labour* (London: Macmillan, 1978); Nancy Chodorow, *The Reproduction of Mothering* (Berkeley: University of California Press, 1978); Keller, *Reflections,* 67–126; Joan Tronto, "Beyond Gender Difference to a Theory of Care," *Signs* 12 (1987):644–63. Tronto identifies the dangers in identifying subjective epistemologies with women only, proposes the relevance of the experience of subordination in producing such ways of knowing, and argues for a more generalized moral theory of caring. Elizabeth Fee connects feminism to other radical epistemologies in "Critiques of Modern Science: The Relationship of Feminism to Other Radical Epistemologies," in Bleier, *Feminist Approaches to Science.*

32. Blackwell, "Why Hygienic Congresses Fail," 2:57, 74, 75; Blackwell, "Influence of Women," 2:12, 19–20, 27–29.

33. Blackwell, "Why Hygienic Congresses Fail," 2:73.

34. See Barbara Rosenkrantz, "Cart before Horse: Theory, Practice, and Professional Image in American Public Health, 1870–1920," *Journal of the History of Medicine and Allied Sciences* 29 (1974): 55–73; Howard Berliner, *A System of Scientific Medicine* (New York: Tavistock, 1985), 76–91; E. Richard Brown, *Rockefeller Medicine Men* (Berkeley: University of California Press, 1979).

35. Eliza Mosher, "The Human in Medicine, Surgery, and Nursing," *Medical Women's Journal* 32 (1925): 117–19; see also Morantz-Sanchez, *Sympathy and Science,* 233–311.

6 / John R. Stilgoe / Plugging Past Reform

In this chapter I focus essentially on Northeast crop raising, but I direct the attention of the reader to the small-scale farmer of the upland South and to livestock enterprises, especially poultry raising. Most analysis of the role of government-backed agricultural science in the reform of agriculture in the post–Civil War Republic focuses on what the scientific establishment did or tried to do; here, I examine what agricultural science avoided.

1. M. C. Betts and W. R. Humphries, *Planning the Farmstead* Farmer's Bulletin 1132 (Washington, D.C.: Government Printing Office, 1920), 5.

2. L. C. Everard, "Science Seeks the Farmer," in *USDA Yearbook, 1920* (Washington, D.C.: Government Printing Office, 1921). See also William L. Bowers, "Country Life Reform, 1900–1920: A Neglected Aspect of Progressive Era History," *Agricultural History* 45 (1971): 211–22.

3. L. C. Everard, "More Complete Knowledge," in *USDA Yearbook, 1920*.

4. "Prints of the Department of Agriculture," in *USDA Yearbook, 1920*.

5. The scale of the publishing effort defies description. The Harvard University collection of experiment station documents is an incredible mass of bound ephemera. The catalogue of the National Agricultural Library boggles the mind by its sheer mass of entries. Who reads 1920s-era farmer's bulletins? Except for historians of agriculture, no one wants outdated information on broody hens or arsenic-based pesticides, and only rarely does anyone disturb the dust of the Harvard collections. I am grateful to Dr. Alan E. Erickson, librarian of Cabot Library, for the extraordinary privilege of spending two years in storage stacks reading experiment station bulletins to find patterns in what is *not* mentioned.

6. Alan I. Marcus, "The Wisdom of the Body Politic: The Changing Nature of Publicly Sponsored American Agricultural Research since the 1830s," *Agricultural History* 62 (1988): 4–26.

7. The best brief overview is Marcus, "Wisdom of the Body Politic," which provides a splendid bibliography as well. The spring 1988 issue of *Agricultural History* is a symposium issue, "Publicly Sponsored Agricultural Research in the United States: Past, Present, and Future," edited by David B. Danbom. Much contemporary research on these issues centers on recent decades. See, for example, Jack Kloppenburg Jr. and Frederick H. Buttel, "Two Blades of Grass: The Contradictions of Agricultural Research as State Intervention," *Research in Political Sociology* 3 (1987): 111–35; Ingolf Vogeler, *The Myth of the Family Farm* (Boulder, Colo.: Westview, 1981); Marty Strange, *The Path Not Taken* (Walthill, N.Y.: Center for Rural Affairs, 1982); and Jack Doyle, *Altered Harvest* (New York: Viking, 1985). Critics of agricultural research by "big agriculture" and "big science" increasingly aim their arguments at the general public, not scholars, or offer alternative scenarios. See, for example, Jim Hightower, *Hard Tomatoes, Hard Times* (Cambridge,

Mass.: Schenkman, 1973); and William H. Friedland and Tim Kappel, *Production or Perish: Changing the Inequalities of Agricultural Research Priorities* (Santa Cruz, Calif.: Project on Social Impact Assessment and Values, 1979).

8. The best retrospective view of attacks on big-agriculture research is David B. Danbom, "Publicly Sponsored Agricultural Research in the United States from an Historical Perspective," in *New Directions for Agriculture and Agricultural Research,* ed. Kenneth A. Dahlberg (Totowa, N.J.: Rowman and Allanheld, 1986). See also Alan I. Marcus, "The Ivory Silo: Farmer–Agricultural College Tensions in the 1870s and 1880s," *Agricultural History* 60 (1986): 22–28; Marcus, *Agricultural Science and the Quest for Legitimacy: Farmers, Agricultural Colleges, and Experiment Stations, 1870–1890* (Ames: Iowa State University Press, 1985).

9. Marcus, "Wisdom of the Body Politic," esp. 21–22.

10. Alan I. Marcus, "From State Chemistry to State Science: The Transformation of the Idea of the Agricultural Experiment Station, 1875–1887," in *Agricultural Scientific Enterprise,* ed. Lawrence Busch and William B. Lacy (Boulder, Colo.: Westview, 1986). See also Charles E. Rosenberg, *No Other Gods: On Science and American Social Thought* (Baltimore: Johns Hopkins University Press, 1976), 173–84. For general studies of the stations, see Norwood Allen Kerr, *The Legacy: A Centennial History of the State Agricultural Experiment Stations, 1887–1987* (Columbia: Missouri Agricultural Experiment Station, 1987); and H. C. Knoblauch, E. M. Law, and W. P. Meyer, *State Agricultural Experiment Stations: A History of Research and Procedure* (Washington, D.C.: USDA, 1962). For a region-specific study, see Karl Quisenberry, "The Dry Land Stations: Their Mission and Their Men," *Agricultural History* 51 (1971): 218–28. Some of the scholarly interest in state experiment stations perhaps derives from the withdrawal of federal support during the Reagan administration, the subsequent reevaluation by states (especially agricultural states) of station roles and support, and worries that land grant universities are losing touch with the social betterment goals of their founders.

11. The term comes from H. M. Railsback, "The Plugger," quoted in W. E. Taylor, *Soil Culture and Modern Farm Methods* (Moline, Ill: Deere and Company, 1924), 2. Taylor uses the term "progressive farmer" too, however, in his list of contributors of ideas to his volume. For him, the term meant what it did to the USDA, but subsequent chapters make clear that his volume aims at the widest possible market for his firm — including pluggers.

12. Taylor, *Soil Culture,* 5.

13. David B. Danbom, *The Resisted Revolution: Urban America and the Industrialization of Agriculture, 1900–1930* (Ames: Iowa State University Press, 1979), is the most penetrating study of farmer unease with science, technological change, and reform. A particularly penetrating study of the 1920s farm crisis, and the ways that USDA-generated statistics and other information still skew historical study of it, is H. Thomas Johnson, *Agricultural Depression in the 1920s: Economic Fact or Statistical Artifact?* (New York: Garland, 1985). Johnson's analysis of the connections between "the Progressive spirit" and "'scientific' statistical analysis" is especially revealing, for it explains how the USDA

might have ignored information that conflicted with its existing notions: "Historians who rely exclusively on USDA bulletins and data to depict agricultural conditions in the 1920s have apparently overlooked this problem" (135).

14. Charles Josiah Galpin, *Rural Life* (1918; New York: Century, 1923), 3.

15. J. Davis, "Posts and Stakes," *New England Farmer* 5 (1853): 208.

16. "Gates vs. Bars," *New England Farmer* 11 (1833): 317.

17. "Gates and Barns," *New England Farmer* 20 (1842): 266.

18. Certainly, many farmers used gates well before the 1850s, as one nostalgic poem makes clear. See "The Old Farm Gate," *New England Farmer* 19 (1841): 256.

19. "Fence Posts," *New England Farmer* 13 (1835): 355.

20. "Improved Plan for Setting Gate and Fence Posts," *New England Farmer* 15 (1836): 12–13.

21. "On the Proper Management of Posts with Reference to Their Durability," *New England Farmer* 15 (1837): 301.

22. "Fences," *Farmer's Cabinet* 4 (1839): 65.

23. "Setting Gate Posts," *New England Farmer* 21 (1842): 89.

24. "Wire Fence," *New England Farmer* 2 (1850): 118; "Iron Posts for Fences," *Country Gentleman* 13 (1859): 27.

25. Benjamin Nott, "Gates," *New England Farmer* 2 (1850): 397–98.

26. W. C. Pinkham, "Farm Gate," *Country Gentleman* 13 (1859): 24.

27. H. L. Brown, "About Gate Making," *Country Gentleman* 13 (1859): 299.

28. H. Hinkley, "Farm Gates," *Country Gentleman* 13 (1859): 313.

29. W. C. Pinkham, "Construction of Farm Gates," *Country Gentleman* 13 (1859): 347.

30. "The Best Gate," *New England Farmer* 14 (1862): 250; "Retrospective Notes," *New England Farmer* 14 (1862): 340.

31. See, for example, E. A. Parks, "A Roller Gate," *New England Farmer* 1 (1867): 375.

32. "Wire Fence," *New England Farmer* 4 (1852): 270. On patent gates, see, for example, "Smith's Vertical Gate," *New England Farmer* 1 (1849): 371; "Baker's Patent Farm Gate," *New England Farmer* 1 (1849): 403.

33. John M. Stahl, "Farm Gates," *American Agriculturist* 43 (1884): 241. See also "A Farm Gate," *American Agriculturist* 39 (1880): 339, which suggests that farmers had increasing access to junk metal to experiment with sliding and rolling rather than swinging gates. The advertisement of manufactured fencing began in earnest in the 1870s and by 1890 was well established. See, for example, "McMullen's Farm Fencing," *American Agriculturist* 49 (1890): 293. I have had the great pleasure of examining the trade catalogues and flyers in the Hagley Museum. Those for the Cyclone Fence Company's Challenge line

offer much material for tracing the transition of manufactured fencing between the 1880s and the 1930s.

34. See, for example, "New England Fences," *Scribner's Monthly* 19 (1880): 502–11, for an early specimen of the genre.

35. Clifton Johnson, *The Farmer's Boy* (New York, 1894), 88, neglects this, for example, as does Herbert Wendell Gleason, "The Old Farm Revisited," *New England Magazine* 22 (1900): 668–80. Now and then, popular periodicals published analytical articles trying to correct the impression made by writers like Johnson and Gleason; see, for example, William H. Brewer, "Is It True that Farming Is 'Declining' in New England?" *New Englander* 52 (1890): 328–42.

36. I have found descriptions of rangeways only in manuscript deeds. See, for example, Liam Hammond to Joshua Reeds, March 31, 1792, Middlesex County Deeds, Registry of Deeds, Cambridge, Massachusetts.

37. "Laying Out Farms," *Illustrated Annual Register of Rural Affairs* 1 (1873): 106–8.

38. James J. H. Gregory, "Observation and Experiment," *Agricultural Societies of Massachusetts for 1859,* ed. Charles L. Flint (Boston, 1860), 1–9. This is a crucial document in the history of agricultural reform.

39. W. E. B., "One-Horse Farmers," *New England Farmer* 15 (1863): 308–9. The impact of the Civil War on one-horse farmers appears to have been great, especially with regard to their acquisition of new machines. See, for example, S. L. Goodale, "The Changes in Farming: Which Have Taken Place and Which Should Be Made," Maine Board of Agriculture, *Seventeenth Annual Report . . . 1872* (Augusta, Maine, 1873), 335–58.

40. "Robinson's Hand Cultivator," *The Plough, the Loom, and the Anvil* 12 (1855): 746–47. Robinson had many colleagues in the business of inventing wheel hoes. See, for example, "The Fairs," in *Massachusetts Board of Agriculture, Fifteenth Annual Report* (Boston, 1868), esp. 287.

41. Information on Allen and Company is derived from the author's collection of company literature assembled over the past several years from booksellers and at farm sales. To purchase Allen and Company literature, one must often acquire the machines.

42. Allen and Company, *Implement Catalogue, 1892* (Philadelphia, 1891), 2, 9, 12, 13.

43. Allen and Company, *Implement Catalogue, 1894* (Philadelphia, 1893), 11.

44. Allen and Company, *Implement Catalogue, 1897* (Philadelphia, 1891), ii–iii. The firm had acquired stiff competition, too; see, for example, *McGee Garden Cultivator* (Rockford, Ill., 1896).

45. Allen and Company, *A Good Garden Is Half of a Good Living* (Philadelphia, 1925), 2. I follow USDA practice here in using *market gardeners* to designate growers selling produce near their farms and *truck farmers* to designate growers shipping (trucking) their produce some distance, often across state lines. In actual usage, no one term designated the small-scale (three-to-fifteen-acre) vegetable grower. For a nationwide attempt

at designation, see W. J. Spillman, "Types of Farming in the United States," in *USDA Yearbook, 1908* (Washington, D.C.: Government Printing Office, 1909), esp. 354. For a local attempt, see John B. Moore et al., "Market Gardening," in *Agriculture of Massachusetts . . . 1870* (Boston, 1871). The authors imply that market gardeners supported wheel hoe inventors: "Clean culture is one of the important things to be practiced in farming; and in this branch of the business it becomes indispensable, and without it there can be no great success" (314).

46. Allen and Company, *Planet Jr. Farm and Garden Implements* (Philadelphia, 1930), 2, 45–46.

47. On truck farming needs, see L. C. Corbett, "Truck Farming in the Atlantic Coast States," in *USDA Yearbook, 1907* (Washington, D.C.: Government Printing Office, 1908); and Fred J. Blair, "Development and Localization of Truck Crops in the United States," in *USDA Yearbook, 1916* (Washington, D.C.: Government Printing Office, 1917). Blair points out the tremendous growth of truck farming and the unwillingness of federal census officials to note it.

48. Rhode Island Agricultural Experiment Station, *Strawberries*, Bulletin 22 (North Kingstown, R.I., 1893), 48–49.

49. See, for example, Allen and Company, *Implement Catalogue, 1892*, 8–11. The Rhode Island experimenters apparently avoided the Planet Jr. wheel hoe developed for strawberry cultivation; see Rhode Island Agricultural Experiment Station, *Strawberries*, 11.

50. Calculating the popularity of wheel hoes is difficult, but see, for example, Winchester-Simmons Co., *Complete (Wholesale) Catalogue* (Chicago (?), 1927), 240–42.

51. See, for example, Allerton S. Cushman, "Information in Regard to Fabricated Wire Fences and Hints to Purchasers," in *USDA Yearbook, 1909* (Washington, D.C.: Government Printing Office, 1910). Some states have long records of evaluating machinery; see, for example, the publications of the Pennsylvania Department of Agriculture, and especially Thomas G. Janvier, "Drills," and Alfred L. Kennedy, "On Mowers and Reapers, and on Corn Harvester," both in *Agriculture of Pennsylvania for 1878* (Harrisburg, Penna., 1879). Both articles cite various manufacturers.

52. Liberty Hyde Bailey, *Blackberries*, Bulletin 99 (Ithaca, N.Y., 1895), 418–19.

53. Liberty Hyde Bailey, *Principles of Vegetable-Gardening* (1901; New York: Macmillan, 1921), 391–96.

54. Cecelia Tichi, *Shifting Gears: Technology, Literature, Culture in Modernist America* (Chapel Hill: University of North Carolina Press, 1987).

55. E. L. D. Seymour, "Economy and the Vegetable Garden," *Country Life* 20 (1911). See also "Market Gardening," *New England Farmer* 75 (1897): 1. This article explicitly discusses the increasing unwillingness of highly competitive market gardeners to share their gardening secrets.

56. See, for example, Wallace S. Moreland, ed., *A Practical Guide to Successful Farming* (Garden City, N.Y.: Halcyon House, 1943), esp. 457–62.

57. W. P. Walker and S. H. DeVault, *Part-Time and Small-Scale Farming in Maryland* (College Park, Md.: Agricultural Experiment Station, 1933), 223.

58. Ray E. Wakeley, *Part-Time and Garden Farming in Iowa* (Ames: Agricultural Experiment Station, 1935), 35–36; Walker and DeVault, *Part-Time and Small-Scale Farming,* 226–27; F. V. Smith and O. G. Lloyd, *Part-Time Farming in Indiana* (Lafayette: Agricultural Experiment Station, 1936), 12–13. For a later study, see *Part-Time Farming in New England* (Amherst: University of Massachusetts Extension Service, 1947). USDA officials now and then noticed—usually in a general, sometimes nostalgic way—the small-scale farmer; see, for example, S. A. Knapp, "Causes of Southern Rural Conditions and the Small Farm as an Important Remedy," in *USDA Yearbook, 1908.*

59. Ervin E. Ewell, "Every Modern Farm an Experiment Station," in *USDA Yearbook, 1897* (Washington, D.C., 1898). How researchers presented agricultural science (not the results) to farmers is beyond the scope of this chapter, but clearly they sometimes presented science as a kind of magic show. See, for example, Dick J. Crosby and B. H. Crocheron, "Community Work in the Rural High School," in *USDA Yearbook, 1910* (Washington, D.C.: Government Printing Office, 1911), esp. 185: "Although alphabetically simple to the chemist, physicist, and soil technologist, the experiments vitally interested the people. Those lamp chimneys and Bunsen flames hypnotically held those in attendance while the talks went on." Agricultural reform as entertainment deserves further study.

60. Gregory, "Observation and Experiment."

61. I base these statements on extensive sampling of state experiment station bulletins; on A. C. True, "Office of Experiment Stations," in *USDA Yearbook, 1897;* and on E. W. Allen, "Some Ways in Which the Department of Agriculture and the Experiment Stations Supplement Each Other," in *USDA Yearbook, 1905* (Washington, D.C.: Government Printing Office, 1906). On extension agents, see Danbom, *Resisted Revolution,* 84–92.

62. Sometime after the Civil War, the term moved from the South, where it applied to tussocks of dark-colored roots projecting above the surfaces of swamps, to the North, perhaps brought back by returning troops. For an introduction to the term, see John Russell Bartlett, ed., *Dictionary of Americanisms* (Boston, 1859), 292. Its racist overtones have made it rarely used, and even elderly New England farmers choose not to use it, which makes difficult their explanations of rock removal in days before bulldozers. Fieldstone can be moved using horses and stoneboats. Niggerheads cannot be moved this way, even though they are not ledge, or living rock. The earthiness, and sometimes brutality, of agricultural English might upset listeners unfamiliar with it, but without some knowledge of it the researcher may ask the wrong sorts of question.

63. See "The Fairs," 289, for quotation.

64. James S. Grinnell, "Hindrances to Successful Farming," in *Massachusetts Board of Agriculture, Thirty-second Annual Report* (Boston, 1885). See also John R. Stilgoe, *Metropolitan Corridor: Railroads and the American Scene* (New Haven: Yale University Press,

1983), 315–33; Howard Fisher Wilson, *Hill Country of Northern New England: Its Social and Economic History* (New York: Columbia University Press, 1936).

65. Institute of Makers of Explosives, *Explosives in Agriculture* (New York, 1931), 62.

66. See John Hensley, "Selling the Blast," *Delaware History* 22 (1986): 99–110, for an excellent analysis of the DuPont effort. The Hagley Library contains a full run of the *Agricultural Blaster* and a wealth of literature on farm uses of dynamite.

67. Institute of Makers of Explosives, *Explosives in Agriculture*, 28. Almost from the beginning of its publication efforts, the institute emphasized the peculiarities of each situation; see, for example, J. R. Mattern, *The Use of Explosives in Making Ditches* (New York: Institute of Makers of Explosives, 1917).

68. H. R. Tolley, *Laying out Fields for Tractor Plowing*, Farmer's Bulletin 1045 (Washington, D.C.: Government Printing Office, 1919).

69. Taylor, *Soil Culture*, 121. Deere and Company, *Operation, Care, and Repair of Farm Machinery* (Moline, Ill., 1936?). Bibliographical information concerning these and other booklets published by implement manufacturers is confused, but the publications clearly deserve study, since they offer insights into complex philosophies (often addressing reform issues) that are markedly different from those presented in agricultural newspapers and government documents. Wayne G. Broehl Jr., *John Deere's Company: A History of Deere & Company and Its Times* (New York: Doubleday, 1984), says little about the firm's educational efforts or its experimental farms, although Deere and Company publications are found in secondhand bookshops in rural America, usually well thumbed and underlined.

70. E. M. Wilson, "Gates Are Cheaper than Bars," *New-England Homestead* 21 (1887): 1.

71. Charles Josiah Galpin, *Rural Social Problems* (New York: Century, 1924), 26–27, 60.

72. The most useful of these journals is *New Farm*. Writers like Wendell Berry and Mark Kramer have introduced much new-old thinking to the general public. See, for example, Wendell Berry, *The Unsettling of America: Culture and Agriculture* (San Francisco: Sierra Club, 1977); and Mark Kramer, *Mother Walter and the Pig Tragedy* (New York: Knopf, 1972). Dismissed in the late 1960s as ridiculous dreamers and in the 1970s as nostalgic fools, "alternative agriculture" farmers became, by the late 1980s, successful growers, often disliked by big agriculture. Farm periodicals like *The Grower* and *Country Folks* now take the arguments of these reformers seriously, but no study yet exists to introduce the movement to serious students of American history and culture.

7 / Roland Marchand and Michael L. Smith / Corporate Science on Display

1. David F. Noble, *America by Design: Science, Technology, and the Rise of Corporate Capitalism* (New York: Knopf, 1977).

2. Arthur Ruel Thompson, "Science Displayed on the Poster," *Printers' Ink Monthly* 19 (1929): 58.

3. John C. Burnham, *How Superstition Won and Science Lost: Popularizing Science and Health in the United States* (New Brunswick: Rutgers University Press, 1987), 64.

4. Leonard S. Reich, *The Making of American Industrial Research: Science and Business at GE and Bell, 1876–1926* (New York: Cambridge University Press, 1985), 5; Charles F. Kettering, "The Head Lamp of Industry," in *Prophet of Progress: Selections from the Speeches of Charles F. Kettering,* ed. T. A. Boyd (New York: Dutton, 1961), 79.

5. Reich, *The Making of American Industrial Research,* 1–2, 4, 8, 206, 240–42; George Wise, *Willis R. Whitney, General Electric, and the Origins of U.S. Industrial Research* (New York: Columbia University Press, 1985), 177, 179, 214; Noble, *America by Design,* 19, 111, 120; Stuart W. Leslie, *Boss Kettering* (New York: Columbia University Press, 1983), 277; Ronald C. Tobey, *The American Ideology of National Science, 1919–1930* (Pittsburgh: University of Pittsburgh Press, 1971), 217, 220, 230. The corporations addressed a public that, as Peter Kuznick points out, drew "no sharp distinction between science and technology." See Peter J. Kuznick, *Beyond the Laboratory: Scientists as Political Activists in 1930s America* (Chicago: University of Chicago Press, 1987), 6.

6. Tobey, *American Ideology,* 97, 101–2, 106–7, 131, 167, 169; James Steel Smith, "The Day of the Popularizers: The 1920s," *South Atlantic Quarterly* 42 (1963): 303.

7. Peter J. Kuznick, "Losing the World of Tomorrow: The Battle over the Presentation of Science at the 1939 World's Fair," *American Quarterly* 46 (September 1994): 342, 360–63. The most famous story of scientific and technological education through fair exhibits was Henry Ford's account of how he had been inspired by an exhibit of small motors for water pumps at the 1893 Columbian Exposition. See David L. Lewis, *The Public Image of Henry Ford: An American Folk Hero and His Company* (Detroit: Wayne State University Press, 1976), 118, 301; and Fred L. Black, Reminiscences (1951), 198, Oral History Collection, Edison Institute, Dearborn, Michigan. For the use of this story to convey the notion of the didactic and inspirational intent of the Ford exhibit at the 1939 World's Fair in New York, see *Saturday Evening Post,* May 6, 1939, 68–69.

8. T. A. Boyd, *Professional Amateur: The Biography of Charles Franklin Kettering* (New York: Dutton, 1957), 217–18; Leslie, *Boss Kettering,* 69. The text of Kettering's 1916 lecture, entitled "Science and the Future Automobile," appears in Boyd, *Prophet of Progress.*

9. John W. Reedy, Oral History, 1, Allen Orth, Oral History, 1, 6, and Paul Garrett, Oral History, 3, 5, in GMI Alumni Foundation Collection of Industrial History, Flint, Michigan (hereafter, GMI); Leslie, *Boss Kettering,* ix–x; Boyd, *Professional Amateur,* 44; Bruce Barton, "Great Things Are Being Done by Men Who Think Boldly," *American Magazine* 97 (March 1924); Bruce Barton to Charles Kettering, January 13, 1923, Walter C. Boynton to C. W. Adams, January 17, 1923, Walter C. Boynton File (1923), GMI; Alfred Swayne to Kettering, May 26, 1923, Kettering to Swayne, May 29, 1923, Alfred Swayne file (1923), Charles F. Kettering Papers, GMI; Paul de Kruif, "Boss Kettering," *Saturday*

Evening Post, July 15, July 29, August 12, and August 26, 1933; Lewis, *The Public Image of Henry Ford,* 289; *Fortune,* 19 (March 1939): 52.

10. Lenox C. Lohr, manuscript of radio speech, July 10, 1929, folder 200, Lenox C. Lohr Papers, University of Illinois, Chicago; Robert W. Rydell, *World of Fairs: The Century-of-Progress Expositions* (Chicago: University of Chicago Press, 1993), 92–101; John M. Staudenmaier "Clean Exhibits, Messy Exhibits: Henry Ford's Technological Easthetic," in *Industrial Society and Its Museums, 1890–1990: Social Aspirations and Cultural Politics,* ed. Brigitte Schroeder-Gudehus (Langhorne, Pa.: Harwood Academic Publishers, n.d.).

11. Lenox Lohr, Address at Luncheon of Advertising Council of Chicago Association of Commerce, January 16, 1930; and Radio Speech, folder 200, Lohr Papers.

12. Press release (on Chicago dinner), May 26, 1934, GM, NY folder (1934), Kettering Papers; Paul Garrett, "The Importance of the Public," speech manuscript (1934), folder 77-7.4-1.13-2, Charles Stewart Mott Papers, GMI; Kuznick, *Beyond the Laboratory,* 24; *Scientific Monthly,* 39 (July 1934), 67–78. The GM dinner, the company's public relations director elatedly reported, had inspired 670 editorials across the nation, all identifying General Motors "with the 'progress' theme." Paul Garrett to Alfred Sloan and C. F. Kettering, June 25, 1934, GM, NY folder (1934).

13. *General Motors World* 13 (July 1934): 6. Kettering had already begun to emphasize the theme of a world as yet unfinished in the talks and interviews on which Barton based his 1924 article. See Barton, "Great Things," 24, 192.

14. Orth, Oral History, 2; *Fortune* 19 (March 1939): 48; "Souvenir Edition, GM Parade of Progress" (1936 brochure), GM, NY folder (1936); Allen Orth to Paul Garrett, March 29, 1935, GM, NY folder (1935); Previews of Progress, Itinerary (typescript, May 3, 1937), GM, NY folder (1937).

15. Orth, Oral History, 1–2; Boyd, *Professional Amateur,* 218; Leslie, *Boss Kettering,* 267; "Souvenir Edition, GM Parade of Progress," Charles Kettering to Alfred P. Sloan Jr., March 19, 1936, all in GM, NY folder (1936); *Daily News* (Miami), February 19, 1936; *Morning Sentinel* (Orlando), March 22, 1936.

16. By the mid-1920s this perpetual motion theory of economic progress had been elaborated by figures like Bruce Barton, Charles Kettering, and research organizer Marcus Alexander, who had molded it into what David Noble describes as a "benevolent circle of prosperity,"—a theory of never-ending material and technological progress. See Noble, *America by Design,* 53, 64. On equivalents of the perpetual motion theory with a focus on consumers, see Daniel Horowitz, *The Morality of Spending: Attitudes toward the Consumer Society in America,. 1975–1940* (Baltimore: Johns Hopkins University Press, 1985), 144.

17. Bruce Barton Speaks, manuscript of speech given October 19, 1925, BBDO Personnel folder, Batten, Barton, Durstine & Osborn Archives, New York City. See also Barton's similar scenario in *Judicious Advertising* 23 (April 1925): 52. Many readers will recognize this story from its revisionist version in the 1981 film, *The Gods Must Be Crazy.*

18. Kettering had publicly set forth nearly all of these ideas during the 1920s. *The New Necessity* simply brought them together in a more sustained manner, with the automobile as the specific case study. Charles Kettering and Allen Orth, *The New Necessity: The Culmination of a Century of Progress in Transportation* (Baltimore: Williams and Wilkins, 1932), 15, 32–33, 57; *Advertising and Selling*, January 9, 1929, 29; "Puts Science on a Production Basis," typescript, enclosure in Walter C. Boynton to C. W. Adams, February 27, 1923, Boynton file (1923); Barton "Great Things," 192–93. See also the penciled comments, apparently by Kettering, in *The Wedge* 33, in Bruce Barton Papers, box 35, Wisconsin State Historical Society. .

19. Leslie, *Boss Kettering,* 258; Kettering and Orth, *The New Necessity,* 33, 113: Alfred P. Sloan Jr., "Science and Industry in the Coming Century," *Scientific Monthly* 39 (July 1934): 69–70. For a good summary of the frontiers-of-science view versus the stagnation view, see Ayer News File, November 5, 1943, 1, N. W. Ayer and Son Archives, New York City.

20. Reedy, Oral History, 6; *Daily News* (Greensboro, N.C.), June 3, 1936; Kettering, "Head Lamp of Industry," 92; W. S. Moore Jr., biweekly report on Texas Centennial Exhibit, series 2, part 2, box 51, E. I. DuPont de Nemours Co. Archives, Hagley Library. Even Kettering found himself protesting that "there is no magic about research. It is just plain hard work." Even if Kettering continued to hope that a "circus of science" might bring about some conversions to the research frame of mind, he recognized the increasing complexity of science and technology and accepted Lohr's warning about avoiding ideas the public could not grasp. He acknowledged that, although the public had once taken an interest in what made cars go, now they wanted to be "unconscious of the mechanism," and he concluded that, since most people could not be told certain technical things that it was necessary to convey, "you have to do something spectacular that will make people want to know the thing you want to tell them." He was not prepared, however, to question whether doing "something spectacular" would actually have that effect. See Charles Kettering, "Can Engineering Principles Be Applied to Advertising?" (May 1927), box 35, Barton Papers; Ed Cray, *Chrome Colossus: General Motors and Its Times* (New York: McGraw-Hill, 1980), 204.

21. *Miami Herald,* February 18, 1936, and February 19, 1936; *Miami Daily News,* February 18, 1936, and February 19, 1936; *Durham Sun,* May 25, 1936; *General Motors World* 15 (February 1936), 4; "Souvenir Edition, GM Parade of Progress"; Marcel Evelyn Chotowski LaFollette, "Authority, Promise, and Expectation: The Images of Science and Scientists in American Popular Magazines, 1910–1955," Ph.D. diss., Indiana University, 1979, 163; Kettering and Orth, *New Necessity,* 3, 6, 10.

22. Leslie, *Boss Kettering,* 15, 55, 308, 335, 339; Boyd, *Professional Amateur,* 28, 70, 77; Kettering, "Head Lamp of Industry," 86; *Journal-American* (New York), August 10, 1941, clipping in file 381 (1: publicity), Norman Bel Geddes Papers, University of Texas, Austin.

23. Rosamond McPherson Young, *Boss Ket: A Life of Charles F. Kettering* (New York: Longmans, Green, 1961), vii; E. I. DuPont de Nemours Company, "Blazing the Trail to New Frontiers through Chemistry," brochure (Wilmington, 1940), n.p; Josephine Young

Case and Everett Needham Case, *Owen D. Young and American Enterprise: A Biography* (Boston: Goidine, 1982), 373; Kettering and Orth, *The New Necessity,* 113; Leslie, *Boss Kettering,* 311. On the image of pioneering, see also Tobey, *American Ideology,* 178. For all of his folksy rhetoric, Kettering also represented the newer generation of scientifically educated engineers and, as Stuart Leslie observes, was eager to take advantage of whatever scientific expertise he could obtain. Moreover, as David Noble reminds us, it was science and management that gave twentieth-century engineers their identity and distinguished them from "mere" mechanics. Leslie surmises that Kettering's pride was deeply wounded by Albert Einstein who reportedly greeted him, on being introduced, with the comment, "Oh, yes, the auto mechanic." Leslie, *Boss Kettering,* 51, 55, 78–79, 98, 138, 318, 327–28; Noble, *America by Design,* 27, 42.

24. Maurice Holland, *Industrial Explorers* (New York: Harper and Brothers, 1929), 16 and passim; John William Ward, "The Meaning of Lindbergh's Flight," *American Quarterly* 10 (Spring 1958): 3–16.

25. "Puts Science on a Production Basis," enclosure, Walter C. Boynton to C. W. Adams, February 27, 1923, Boynton file (1923); Kettering and Orth, *New Necessity,* 48, 53–58.

26. Julian Lewis Watkins, *The One Hundred Greatest Advertisements: Who Wrote Them and What They Did,* 2d ed. (New York: Dover, 1959), 123; David E. Nye, *Image Worlds: Corporate Identities at General Electric, 1890–1930* (Cambridge: MIT Press, 1985), 106–7.

27. Burnham, *How Superstition Won,* 7, 169; Kuznick, *Beyond the Laboratory,* 42; LaFollette, "Authority, Promise, and Expectation" 97–99; Marcel C. LaFollette, *Making Science Our Own: Public Images of Science, 1910–1955* (Chicago: University of Chicago Press, 1990), 71–72, 122. LaFollette also notes how frequently magazine writers used such phrases as "unremitting toil," "years of patient, tireless effort," and "personal sacrifice" to describe scientists. On Kettering's contempt for leisure and his inability to understand "that part of the twenty-four hours a day could be spent in other ways than working," see Young, *Boss Ket,* 150, 198. For the origins of the fear that technology might lead to enervation and "enjoyment without the sweat," see John Kasson, *Civilizing the Machine: Technology and Republican Values in America, 1778–1900* (New York: Grossman, 1976), 124–25.

28. Boyd, *Professional Amateur,* 55–56, 73, 94, 113, 120, 160; Young, *Boss Ket,* 65–67, 106, 150. See also de Kruif, "Boss Kettering."

29. Daniel J. Kevles, *The Physicists: The History of a Scientific Community in Modern America* (New York: Knopf, 1978), 205–6; LaFollette, "Authority, Promise, and Expectation," 180–81, 193.

30. Stuart and Elizabeth Ewen, *Channels of Desire: Mass Images and the Shaping of American Consciousness* (New York: McGraw-Hill, 1982), 149, say that the female was "to be seen as one who bewitches and plays on the benevolent weakness of a man who—in terms of his own patterns of consumption—remains uncorrupted."

31. Leslie, *Boss Kettering,* 243. In *American Ideology* (923) Ronald Tobey effectively

summarizes the self-image of dirty, hard work and contempt for leisure that scientists and their popularizers cultivated. Edwin Slosson, the energetic editor of *Science Service,* Tobey notes, always depicted scientists as "doing something . . . involved in hard work." For the "pioneers of science," research "was a serious and solemn thing." It was "toil" and "drudgery." Although the result of their work was to relieve housewives of "drudgery," scientists and engineers had no desire to see themselves relieved of challenges that involved hard work.

32. Wise, *Willis R. Whitney,* 233; BDO, "Report and Recommendations, 1926," 16, GE Archives, Fairfield, Conn.; Nye, *Image Worlds,* 124–32. In another questionnaire directed to 365 college men and women, 194 respondents identified GE as a corporation that conducted scientific research compared to only 43 who named Western Electric, 20 who named DuPont, and 15 who named AT&T. P. L. Thomson to F. W. Jewett, February 6, 1925, F. W. Jewett Papers, AT&T Archives, Warren, N.J.

33. Wise, *Willis R. Whitney,* 214; *Saturday Evening Post,* March 28, 1925, 82, and September 6, 1930, 62; Bruce Barton to Gerard Swope, November 20, 1922, April 3, 1924, and Barton, "A Memo for Mr. Swope," 3, enclosure, Barton to Swope, September 27, 1922, Swope files, GE Archives; Edward Gibbons, *Floyd Gibbons: Your Headline Hunter* (New York, 1953), 218–19, 224–25.

34. *Fortune* 19 (March 1939): 109; Gibbons, *Floyd Gibbons,* 224–25; "Knowing More about GE" (1932), 5150–51, Hammond files, GE Library, Schenectady, N.Y.; Lawrence A. Hawkins, *Adventure into the Unknown: The First Fifty Years of the General Electric Research Laboratory* (New York, 1950), 78; Lawrence A. Hawkins, "Reminiscences" (1951), Oral History Research Project, Columbia University, 25; Wise, *Willis R. Whitney,* 284, 288; Bruce Barton to Gerard Swope, September 8, 1930, Robert Updegraff to Barton, September 5, 1930, T. K. Quinn to Porter, Quinn, et al., September 22, 1930, Swope files.

35. *Saturday Evening Post,* September 6, September 20, October 4, 1930; Martin Rice to Gerard Swope, September 27, 1930, Katherine Woods to Swope, April 23, 1935, Swope files; GE, "Our Friends Call It 'The House of Magic'" (1933), Hammond files L2024.

36. *GE Monogram,* April 1933, 12, and June 1933, 1–3; C. H. Lang to Gerard Swope, September 14, 1935, Swope files; "1934 Report on the Fair," acc. 1109, box 7, Ford Motor Company Archives, Edison Institute, Dearborn, Mich.; "Topics for Public Relations Committee Meeting, November 12, 1932," GM, NY folder (1932); "An Experiment in Advertising and Publicity" (1936), pamphlet file, GM Public Relations Library, Detroit; Press release, May 3, 1934, Century of Progress Publicity Department, untitled typescript, May 8, 1934, untitled description of House of Magic, June 22, 1934, Century of Progress Papers, 1–6177, University of Illinois, Chicago.

37. On this point, see Jeffrey L. Meikle, *Twentieth Century Limited: Industrial Design in America, 1925–1939* (Philadelphia, 1979), 199.

38. Dorothy Nelkin, *Selling Science: How the Press Covers Science and Technology* (New York, 1987), 78; Burnham, *How Superstition Won,* 6–7, 12, 215; N. W. Ayer and Son,

"Every-day Magic" (1928), box 15-51, Ayer Archives, New York; LaFollette, "Authority, Promise, and Expectation," 12, 249.

39. "1934 Report on the Fair," 12; *Ad-vents: Weekly News of the Greater Buffalo Advertising Club,* August 15, 1939, 1939 folder, Kettering Papers; Martin Dodge to Walter Dorwin Teague, September 17, 1937, reel 16:23A, Walter Dorwin Teague Papers, George Arents Library, Syracuse Library.

40. Roland Marchand, "The Designers Go to the Fair: Walter Dorwin Teague and the Professionalization of Corporate Industrial Exhibits, 1933–1940," *Design Issues* 7 (Spring 1991): 4–17.

41. "The Bell Telephone Exhibit at the New York World's Fair, 1939–40," box 175.08.02, AT&T Archives; "The Middleton Family Goes to the Fair," film, Prelinger and Associates, New York; "Selecting World's Fair Exhibit," 23, box 1061, AT&T Archives; AMI Distribution Company, "Synchronizing and Amplifying Systems," file 381 (specifications, building), Bel Geddes Papers; Kuznick, "Science Populizers," 30; *Business Week,* November 4, 1939, 27.

42. "Description of the Chrysler Motors Exhibits at the New York World's Fair, 1939," 330:3, and Chrysler press release, April 27, 1939, vol. 334, T. J. Ross Scrapbooks, Ivy Lee Papers, Princeton University.

43. Photograph 72.341.76, DuPont Photographic Archives, Hagley Library; DuPont, "The 1940 Wonder World of Chemistry," box 1005, New York World's Fair Collection, Library (hereafter, NYPL). As Jeffrey Meikle convincingly argues, the wonders-of-science approach did not prove to be the optimum marketing technique nor did it cultivate the most favorable attitudes toward the corporation. Visitors to the DuPont exhibit clearly preferred that the company's exciting new product, nylon stockings, "be domesticated" and "rendered meaningful to everyday life" than mystified as an aspect of "chemical utopianism." See Jeffrey L. Meikle, *American Plastic: A Cultural History* (New Brunswick: Rutgers University Press, 1995), 142–45. Perhaps the company displayed better judgment when it stretched "the legitimating power of the museum . . . to the limit" by staging a version of its World's Fair fashion show in connection with its scientific exhibit within the sanctified precincts of New York's Museum of Science. See David Rhees, "Corporate Advertising, Public Relations, and Popular Exhibits: The Case of DuPont," in Schroeder-Gudehus, *Industrial Society,* 69.

44. Walter Dorwin Teague to William H. Hart, June 22, 1936, Teague to Carl J. Hasbrouck, June 22, 1936, reel 16:21A, Teague Papers; Oversize Scrapbook, acc. 77.242, DuPont Photographic Archives; "Young Henry Ford Went to the Fair," Ayer and Son, "Ford Cycle of Production, 1940, Revised," enclosure to H. L. McClinton to Fred Black, June 18, 1940, acc. 544, box 21, Henry Billings, "Report on Mural Decorations for Entrance Hall," acc. 544, box 1, press release May 13, 1939, acc. 544, box 14, Ford Archives; *Advertising and Selling,* February 1939, 22–23.

45. Billings, "Report on Mural Decorations"; Edward Mabley, "Transportation Pag-

eant," October 19, 1937, Walter D. Teague, "Ford Exposition, 1940," reel 16:23A, Teague Papers.

46. Meikle, *Twentieth Century Limited*, 201–2; Roland Marchand, "The Designers Go to the Fair II: Norman Bel Geddes, the General Motors 'Futurama,' and the Visit to the Factory Transformed," *Design Issues* 8 (Spring 1992): 23–40.

47. Norman Bel Geddes, "Description of the General Motors Building and Exhibit for the New York World's Fair," September 8, 1939, file 381, Bel Geddes, "Magic Motorways," file 384, and the *Sun* (New York), April 29, 1939, clipping (1, publicity), file 381, Bel Geddes Papers.

48. "1939 Plan for the Fair?—Design Board Schemes Rejected," *Architectural Record*, January 1961, 23 ("The Fair Corporation rejected the [architects'] favored plan and adopted essentially the same layout as the 1939 Fair to make use of existing utilities"). See also Robert A. Caro, *The Power Broker: Robert Moses and the Fall of New York* (New York: Knopf, 1974), 1092.

49. "Case Study No. 905," *Public Relations News*, n.d., box 405, 1964–65 New York World's Fair Papers, NYPL.

50. *Advertising Age*, September 3, 1962; "Your Day at the New York World's Fair, 1964–1965," Hagley Library.

51. "World's Fair Proposed Radio Commercial," J. Walter Thompson Agency, May 17, 1965, radio commercial for February 11, 1964 special World's Fair edition of *Look* magazine, box 406, NYPL.

52. David Brinkley, "Taking the Tour," *Official Preview: New York World's Fair, 1964/1965* (New York: Time Inc., 1963), n.p.; William Doll, cited in ibid.

53. J. E. (Jiggs) Weldy, "Using Fairs Effectively: The Show, Flow, Glow, and Dough Elements of World's Fair Participation," in *Key Facts for Advertisers on the NYWF 1964–1965: Third Report* (New York: Association of National Advertisers, 1962), 18. Weldy warned that "if you have a lot of billboards and attractive things for people to read, you must expect that the most interested readers will be your own company people." Fair exhibits, however, could contribute to internal goals among one's own "company people": "We also feel that we can use this [fair promotion] as bait, if you please, to stimulate increased activity and action amongst all our various departments."

54. James Gardner, "Exhibit Designs and Techniques that Attract the General Public," in *Key Facts for Advertisers*, 25–26. "When they enter that show," Gardner observed of fair audiences, "they are tuned in to a much higher pitch of critical appreciation than they normally are. In fact, let's face it, most people are tuned in to no pitch of critical appreciation."

55. Stanley Appelbaum, *The New York World's Fair, 1939/1940* (New York: Dover, 1977), 68–69; Brinkley, "Taking the Tour," n.p. The Electric Power and Light Building also exhibited mechanical hens that sang, "We chickens all have to cluck with pride / about our farms electrified."

252 / *Notes to Pages 170–175*

56. Caro, *Power Broker*, 1084–114. See also Marc H. Miller, "Something for Everyone: Robert Moses and the Fair," in *Remembering the Future: The New York World's Fair from 1939 to 1964* (New York: Queens Museum, 1989), 62–65.

57. "Breakfast with Dorothy [Kilgallen] and Dick [Kollmer]," Radio-TV Reports, June 6, 1962, WOR radio, box 406, NYPL. For the Cold War's impact on the fair, see Michael L. Smith, "Making Time: Representations of Technology at the 1964 Fair," in *The Power of Culture: Critical Essays in American History*, ed. Richard W. Fox and T. J. Jackson Lears (Chicago: University of Chicago Press, 1993), 223–44.

58. *Fair News*, November 22, 1963. See also Rosemarie Haag Bletter, "The 'Laissez-Fair,' Good Taste, and Money Trees: Architecture at the Fair," in *Remembering the Future*, 126–28.

59. "Corporate" radio program (WQXR), William H. Ottley to Frank C. Kline, May 20, 1965, box 406, NYPL. GE ran spot commercials on several radio stations, promoting the fair and Progressland.

60. "Science at the Fair," box 141, NYPL. See also Sheldon Reaven, "New Frontiers: Science and Technology at the Fair," in *Remembering the Future*, 75–103.

61. Patricia Leighton-Brown, "Fifty Years after the Fair," *New York Times*, March 2, 1989.

62. "Science at the Fair." See also Daniel Cohen, "Preview of Disney's World's Fair Shows," *Science Digest*, December 1963, 9–15; and "*Look*'s Guide to the N.Y. World's Fair," *Look*, February 11, 1964.

63. In his report on the opening of the fair, WBS-TV reporter Douglas Edwards spoke of the "GM soothsayers" who "trust that progress will solve the problems of city traffic just as the superhighways they predicted made road travel easier in our own time" ("Late News," Radio-TV Reports, April 8, 1964, box 406 NYPL).

64. General Motors, *Let's Go to the Fair and Futurama: NYWF, 1964–1965;* Joyce Martin, "Transportation Area Exhibits" (press release), April 21, 1965, box 393, NYPL.

65. In addition to designing Ford's Magic Skyway and GE's Carousel of Progress, Disney and his staff created the Pepsi-Cola exhibit, "It's a Small World," and an audio-animatronic Abraham Lincoln for the Illinois pavilion. The Carousel, "It's a Small World," and Lincoln (now joined by other presidents) now reside at Disney World.

66. Charles Ridgeway, interview with author, August 23, 1988.

67. Disney had made the whimsical depiction of the past a regular staple of his work long before EPCOT. In the mid-1950s, he produced a series of television programs on space travel that included a slapstick animated history of rocketry. "The history," Disney said during a script meeting, "might be a good way to work in a lot of your laughs." See Brian Horrigan, "Popular Culture and Visions of the Future in Space, 1901–2001," in *New Perspectives on Technology and American Culture*, ed. Bruce Sinclair (Philadelphia: American Philosophical Society, 1986), 61.

68. See, for example, Bertram Wolfe, "The Hidden Agenda," in *Nuclear Power: Both Sides*, ed. Michio Kaku and Jennifer Trainer (New York: Norton, 1982), 240–43; and Theos J. Thompson, "Improving the Quality of Life—Can Plowshare Help?" *Symposium on Engineering with Nuclear Explosives, January 14–15, 1970* (AEC:CONF-700707), 1:104.

69. General Motors, "Let's Go to the Fair."

70. Brown, "Fifty Years after the Fair."

8 / James T. Kloppenberg / Why History Matters to Political Theory

1. I do not mean to suggest that analytic philosophy and objective social science are dead. For better or worse, they obviously remain very much alive.

2. Thomas S. Kuhn, *The Structure of Scientific Revolution* (Chicago: University of Chicago Press, 1962).

3. The best study of Dilthey in English is Michael Ermarth, *Wilhelm Dilthey: The Critique of Historical Reason* (Chicago: University of Chicago Press, 1978).

4. Peter Novick, *That Noble Dream: The "Objectivity Question" and the American Historical Profession*. Ideas in Context (Cambridge: Cambridge University Press, 1988). For overviews, see Richard Bernstein, *The Restructuring of Social and Political Theory* (New York: Harcourt, Brace, Jovanovich, 1976); Fred R. Dallmayr and Thomas A. McCarthy, eds., *Understanding and Social Inquiry* (Notre Dame: University of Notre Dame Press, 1977); Quentin Skinner, ed., *The Return of Grand Theory in the Human Sciences* (Cambridge: Cambridge University Press, 1985).

5. Don Herzog, *Without Foundations: Justification in Political Theory* (Ithaca: Cornell University Press, 1985), 20; Joseph Raz, *The Morality of Freedom* (Oxford: Clarendon, 1986), 3; Richard Flathman, *The Philosophy and Politics of Freedom* (Chicago: University of Chicago Press, 1987). Flathman elaborates his position in his collection of essays *Toward a Liberalism* (Ithaca: Cornell University Press, 1989).

6. John Rawls, "Kantian Constructivism in Moral Theory," *Journal of Philosophy* 77 (1980): 515–72; John Rawls, "Justice as Fairness: Political, Not Metaphysical," *Philosophy and Public Affairs* 14 (1985): 223–51; the quotations are from 228, 238, and 249. For Rawls' elaboration of his revised position, see Rawls, *Political Liberalism* (New York: Columbia University Press, 1993). On Rawls's constructivism, see Richard Rorty, "The Priority of Democracy to Philosophy," in *The Virginia Statute for Religious Freedom: Its Evolution and Consequences in American History*, ed. Merrill D. Peterson and Robert C. Vaughan (Cambridge: Cambridge University Press, 1988). For a critical assessment of the adequacy of Rawls's historicist turn, see Patrick Neal, "Justice as Fairness: Political or Metaphysical?" *Political Theory* 18 (1990): 24–50; for a defense of Rawls, cf. Amy Gutmann, "The Central Role of Rawls's Theory," *Dissent* (1989): 338–42.

7. Amy Gutmann, "Communitarian Critics of Liberalism," *Philosophy and Public Affairs* 14 (1985): 308–22; Stephen Holmes, "The Community Trap," *New Republic*, Nov. 28,

1988, 24–28; Rorty, "Priority of Democracy," 272; Jeffrey Stout, *Ethics after Babel: The Languages of Morals and Their Discontents* (Boston: Beacon, 1988), 229.

8. Stephen Holmes, *The Anatomy of Antiliberalism* (Cambridge: Harvard University Press, 1993); Holmes, *Passions and Constraint: On the Theory of Liberal Democracy* (Cambridge: Harvard University Press, 1995); the quotation is from xii.

9. Michael Walzer, *Spheres of Justice: A Defense of Pluralism and Equality* (New York: Basic Books, 1983), 6. Introducing a series of state-of-the-art essays in a recent issue of *Dissent*, Walzer notes "a series of convergences: between liberals and socialists, Marxists and pluralists, defenders of community and defenders of individual autonomy." In his estimation, "a renewed sense of the value of liberalism and social democracy is their single most important cause." While I too would emphasize that factor in explaining this reorientation, the contributors of several of the essays, including notably Alan Ryan on communitarianism, Don Herzog on liberalism, David Plotke on Marxism, and Sanford Levinson on critical legal studies, also note the importance of an increased awareness of historicity and resistance to foundationalism in accounting for the convergences Walzer stresses. See Walzer "The State of Political Theory," *Dissent* (1989): 337–70; Walzer, *Thick and Thin: Moral Argument at Home and Abroad* (Notre Dame: University of Notre Dame Press, 1994); the quotation is from 27.

10. William Sullivan's contributions to Robert N. Bellah, Richard Madsen, William M. Sullivan, Ann Swidler, and Steven M. Tipton, *Habits of the Heart: Individualism and Commitment in American Life* (Berkeley: University of California Press, 1986), are especially apparent in chapters 2 and 10; compare his *Reconstructing Public Philosophy* (Berkeley: University of California Press, 1982). Charles Taylor, *Philosophy and the Human Sciences: Philosophical Papers* (Cambridge: Cambridge University Press, 1985), 2:312–13; see also Taylor, *Sources of the Self: The Making of the Modern Identity* (Cambridge: Harvard University Press, 1989); the quotation is from 520. Compare William Galston *Liberal Purposes: Goods, Virtues, and Diversity in the Liberal State* (Cambridge: Cambridge University Press, 1991), who argues that liberalism itself, independent of the religious convictions that animated its early champions, contains adequate resources to sustain a conception of the good and virtuous life. William E. Connolly, *Politics and Ambiguity* (Madison: University of Wisconsin Press, 1987); Richard Bernstein, *Beyond Objectivism and Relativism* (Philadelphia: University of Pennsylvania Press, 1983); Richard Bernstein, *Habermas and Modernity* (Cambridge: Polity, 1985); Thomas McCarthy, *The Critical Theory of Jürgen Habermas*, 2d ed. (Cambridge: MIT Press, 1985). I am aware of the difficulties involved in grouping disparate thinkers under the unwieldy labels *liberal* and *communitarian*. On this division, see John R. Wallach, "Liberals, Communitarians, and the Tasks of Political Theory," *Political Theory* 15 (1987): 581–611; and cf. Richard Bernstein, "One Step Forward, Two Steps Back: Richard Rorty on Liberal Democracy and Philosophy," *Political Theory* 15 (1987): 561, nn. 12, 14.

11. J. Donald Moon, review of Flathman, *The Philosophy and Politics of Freedom*, in *Political Theory* 16 (1988): 650–54; Stout, *Ethics after Babel*, 227; William A. Galston, "Community, Democracy, Philosophy: The Political Thought of Michael Walzer," *Politi-*

cal Theory 17 (1989): 119–30; Bernstein, introduction to *Habermas and Modernity*, 1–32; John B. Thompson and David Held, eds., *Habermas: Critical Debates* (Cambridge: MIT Press, 1982).

12. Compare Michael Sandel, *Liberalism and the Limits of Justice* (Cambridge: Cambridge University Press, 1982); and Sandel, *Democracy's Discontent: America in Search of a Public Philosophy* (Cambridge: Harvard University Press, 1996); the quotation is from 28.

13. Richard Rorty, ed., *The Linguistic Turn: Recent Essays in Philosophical Method* (Chicago: University of Chicago Press, 1967); Richard Rorty, *Philosophy and the Mirror of Nature* (Princeton: Princeton University Press, 1979), 392; Richard Rorty, *The Consequences of Pragmatism* (Minneapolis: University of Minnesota Press, 1982), xl, xliii. I discuss Rorty's work in greater detail, and place it in the context of the broader revival of pragmatism in recent American thought, in Kloppenberg, "Pragmatism: An Old Name for Some New Ways of Thinking?" *Journal of American History* 83 (1996): 100–138. The phrase "historicist undoing" is from Ian Hacking, "Two Kinds of 'New Historicism' for Philosophers," in his *History and. . . : Histories within the Human Sciences* (Charlottesville: University Press of Virginia, 1995).

14. Rorty, *Consequences*, xviii, 160, 161.

15. Richard Rorty, *Objectivity, Relativism, and Truth* (Cambridge: Cambridge University Press, 1991), 8. Rorty's essay "Solidarity or Objectivity," in Rajchman and West, *Post-Analytic Philosophy*, together with many others Rorty wrote in the 1980s, are collected in two volumes published in 1991 by Cambridge University Press: *Objectivity, Relativism, and Truth* and *Essays on Heidegger and Others*. On the incoherence of the problem of relativism from a pragmatic perspective, see especially Donald Davidson, "On the Very Idea of a Conceptual Scheme," in Rajchman and West, *Post-Analytic Philosophy*.

16. Richard Rorty, *Contingency, Irony, and Solidarity* (Cambridge: Cambridge University Press, 1989), xvi, 53, 189–90. See also Rorty, *Consequences*, 166.

17. Georg Iggers, introduction to Leopold von Ranke, *The Theory and Practice of History*, discussed in Novick, *Noble Dream*, 27–28.

18. James T. Kloppenberg, *Uncertain Victory: Social Democracy and Progressivism in European and American Thought, 1870–1920* (New York: Oxford University Press, 1986), 107–14.

19. The quotation from Wilhelm Dilthey is from his *Selected Writings*, ed. and trans. H. P. Rickman (Cambridge: Cambridge University Press, 1976), 120. The writings of John Dunn and Quentin Skinner were of course seminal in resurrecting the historical analysis of political ideas. Among recent examples emphasizing the importance of the historicist turn in political theory, see Alasdair MacIntyre, "The Indispensability of Political Theory," in *The Nature of Political Theory*, ed. David Miller and Larry Siedentop (Oxford: Oxford University Press, 1983); Herzog, *Without Foundations*, 218–43; Don Herzog, "Approaching the Constitution," *Ethics* 99 (1988): 147–54; Stout, *Ethics after Babel*, 47, 72–73, 120. For a more detailed discussion of historicism, see James T. Kloppenberg, "Objec-

tivity and Historicism: A Century of American Historical Writing," *American Historical Review* 94 (1989): 1011–30.

20. A partial list of recent works demonstrating the inadequacy of attempts to characterize American political thought as straightforwardly "liberal" or "republican" includes Jack N. Rakove, *Original Meanings: Politics and Ideas in the Making of the Constitution* (New York: Knopf, 1996); and Donald S. Lutz, *A Preface to American Political Theory* (Lawrence: University Press of Kansas, 1992), both of whom emphasize the role of Americans' experience in shaping our characteristic form of political expression, constitution writing and institution building; Robert Shalhope, *The Roots of Democracy: American Thought and Culture, 1760–1800* (Boston: Twayne, 1990); and Lance Banning, *The Sacred Fire of Liberty: James Madison and the Founding of the Federal Republic* (Ithaca: Cornell University Press, 1995), both of whom see the blending of religious, republican, and liberal ideas in the founding more clearly than do Gordon Wood, who discounts religion entirely in *The Radicalism of the American Republic* (New York: Knopf, 1991); and J. C. D. Clark, who treats the Revolution as the last war of religion waged by anti-Catholic zealots in *The Language of Liberty, 1660–1832: Political Discourse and Social Dynamics in the Anglo-American World* (Cambridge: Cambridge University Press, 1994).

Studies acknowledging that an interest in securing individuals' supposedly "liberal" property rights was not incompatible with a supposedly "republican" interest in justice include Joyce Appleby, *Liberalism and Republicanism in the Historical Imagination* (Cambridge: Harvard University Press, 1992); Isaac Kramnick, *Republicanism and Bourgeois Radicalism: Political Ideology in Late Eighteenth-Century England and America* (Ithaca: Cornell University Press, 1990); Richard F. Teichgraeber III, *Sublime Thoughts/Penny Wisdom: Situating Emerson and Thoreau in the American Market* (Baltimore: Johns Hopkins University Press, 1995); Richard Ellis, *American Political Cultures* (New York: Oxford University Press, 1993); and especially William J. Novak, *The People's Welfare: Law and Regulation in Nineteenth-Century America* (Chapel Hill: University of North Carolina Press, 1996), which demonstrates brilliantly that nineteenth-century America's supposedly liberal courts continued to embrace, and enforce, an ideal of the public good that drew on various sources and incorporated ideas of virtue, freedom, and social responsibility.

For my own arguments on the multiple sources and complex varieties of American political thought and behavior, which cannot be reduced to a conflict between "liberalism" and "republicanism" without serious distortion, see James T. Kloppenburg, "The Virtues of Liberalism: Christianity, Republicanism, and Ethics in American Political Discourse," *Journal of American History* 74 (1987): 9–33; Kloppenberg, "Religion and Politics in American Thought since the Enlightenment," in *Knowledge and Belief in America: Enlightenment Traditions and Modern Religious Thought,* ed. William M. Shea and Peter A. Huff (New York: Cambridge University Press, 1995); Kloppenberg, "Republicanism in American History and Historiography," *La Revue Tocqueville/The Tocqueville Review* 13 (1992): 119–36; and Kloppenberg, "Political Ideas and Movements," in vol. 1, *Encyclopedia of the United States in the Twentieth Century,* ed. Stanley Kutler (New York: Scribners, 1995).

21. The problems involved in simplifying history for political purposes rather than deriving from it an understanding of the complexity of the past is equally apparent in the recent surge of interest in legal history and the diverse appropriations of historical evidence by lawyers across the political spectrum. On this issue, see Laura Kalman, *The Strange Career of Legal Liberalism* (New Haven: Yale University Press, 1996); and James T. Kloppenberg, "Deliberative Democracy and Judicial Supremacy," *Law and History Review* 13 (1995): 393–411.

22. Thomas Haskell, "The Curious Persistence of Rights Talk in the 'Age of Interpretation,'" *Journal of American History* 74 (1987): 984–1012; Amy Gutmann, ed., *Democracy and the Welfare State* (Princeton: Princeton University Press, 1988), especially the essays by Michael Walzer, J. Donald Moon, and Jon Elster.

23. Daniel Rodgers, *Contested Truths* (New York: Basic Books, 1987); Bernstein, "One Step Forward," 552; Richard Rorty, "Postmodern Bourgeois Liberalism," in *Hermeneutics and Praxis,* ed. Robert Hollinger (Notre Dame: University of Notre Dame Press, 1985); MacIntyre, "Indispensability of Political Theory"; Bellah, *Habits of the Heart;* Walzer, *Spheres of Justice;* Stout, *Ethics after Babel;* Herzog, *Without Foundations;* Nancy Fraser, "Solidarity or Singularity? Richard Rorty between Romanticism and Technocracy," in Fraser, *Unruly Practices* (Minneapolis: Minnesota University Press, 1989); Richard J. Bernstein, "One Step Forward, Two Steps Backward: Rorty on Liberal Democracy and Philosophy," *Political Theory* 15 (1987): 538–63; reprinted in Bernstein, *The New Constellation: The Ethical-Political Horizons of Modernity/Postmodernity* (Cambridge: MIT Press, 1992); the quotation is from 245. See also Rorty's response to Bernstein, "Thugs and Theorists," *Political Theory* 15 (1987): 564–80.

24. Rorty, "Thugs and Theorists," 565, 571, 577, n. 19; Rorty, *Contingency,* 181–82.

25. Walter Rauschenbusch, *Christianity and the Social Crisis* (1907; New York: Harper-Torchbooks, 1964), 408, 252.

26. Compare James T. Kloppenberg, "Democracy and Disenchantment: From Weber and Dewey to Habermas and Rorty," in *Modernist Impulses in the Human Sciences, 1870–1930,* ed. Dorothy Ross (Baltimore: Johns Hopkins University Press, 1994); and Richard Rorty, "Dewey between Hegel and Darwin," in ibid.

27. Rorty, *Contingency,* 67–68. On the similarities between Habermas and Dewey, see Bernstein, *Philosophical Profiles,* 91; and Rorty, "Thugs and Theorists," 580, n. 31.

28. Peter Dews, ed., *Habermas, Autonomy and Solidarity: Interviews* (London: Verso, 1986), 91, quoted in Rorty, "Thugs and Theorists," 565, 575, n. 6. For Habermas's conception of the relation between cognitive-instrumental rationality and the life world, see Habermas, *The Theory of Communicative Action,* trans. Thomas McCarthy, 2 vols. (Boston: Beacon, 1984, 1987).

29. Rorty, *Contingency,* 68.

30. John Dewey, "Liberalism and Social Action: The Page-Barbour Lectures," in Dewey, *The Later Works, 1925–1953,* vol. 11, *1935–1937,* ed. Jo Ann Boydston (Carbondale:

Southern Illinois University Press, 1987), 27. Cf. Dewey, "Art as Experience," in ibid., vol. 10, *1934*, 350, quoted in Rorty, *Contingency*, 69.

31. Stephen Toulmin, "The Recovery of Practical Philosophy," *American Scholar* 57 (1988): 337–52; and his elaboration of this argument in Stephen Toulmin, *Cosmopolis: The Hidden Agenda of Modernity* (New York: Free Press, 1990).

32. Thomas Haskell, "Capitalism and the Origins of the Humanitarian Sensibility," pts. 1 and 2, *American Historical Review* 90 (1985): 339–61, 457–66; Thomas Haskell, "Convention and Hegemonic Interest in the Debate over Antislavery: A Reply to Davis and Ashworth," *American Historical Review* 92 (1987): 829–79. The contributions of Haskell, David Brion Davis, and John Ashworth to this exchange on the sources of anti-slavery sentiment, together with a valuable introduction by Thomas Bender, are available in Thomas Bender, ed., *The Antislavery Debate: Capitalism and Abolitionism as a Problem of Historical Interpretation* (Berkeley: University of California Press, 1992). For further illumination of the diverse sources of antebellum reformist sentiment, see the brilliant essay by Elizabeth B. Clark, " 'The Sacred Rights of the Weak': Pain, Sympathy, and the Culture of Individual Rights in Antebellum America," *Journal of American History* 82 (1995): 463–93; and Daniel Walker Howe, *Virtue, Passion, and Politics: The Construction of the Self in American Thought, from Jonathan Edwards to Abraham Lincoln and Beyond* (Cambridge: Harvard University Press, forthcoming).

33. Rorty, *Contingency*, 196; Rorty, *Consequences*, 162. On the question of dualism in Rorty's recent writings, see Stout's parallel formulation in *Ethics after Babel*, 241, 261–63, 292. There is a paradoxical and incongruous relation between the distinction Rorty now insists upon between the private and public spheres and his stiff resistance to the genre distinction between what Habermas calls the "world-disclosing" and "problem-solving" capacities of "art and literature on the one hand, and science, morality, and law on the other." When Rorty offers historical knowledge of our tradition as the source of the only standard of comparison we can offer to persuade others of the value of our values, he seems to rely on precisely the genre distinction he refuses to allow Habermas to make. It is, after all, not the quality of the literature but the quality of the lives our culture makes possible that Rorty offers as evidence of the superiority—for us—of liberal democracy. Cf. Jürgen Habermas, "Excursus on Leveling the Genre Distinction between Philosophy and Literature," in Habermas, *The Philosophical Discourse of Modernity: Twelve Lectures*, trans. Frederick Lawrence (Cambridge: MIT Press, 1986); Rorty, "Thugs and Theorists," 579, n. 26; Rorty, "Solidarity and Objectivity," 11; Rorty, *Contingency*, 53, 84–85. Some of Rorty's most recent writings, in which he criticizes the academic left for ignoring poverty and inequality and wallowing in questions of identity irrelevant to the most urgent problems facing Americans today, signal a refreshing willingness to address questions of politics that I find consistent with Dewey's own stance. So too are Rorty's recent admission of the contested terrain of "our tradition" and his recommendation that we can find in American history (rather than in contemporary political philosophy or critical theory) arguments on behalf of equality and justice useful for those engaged in the struggles

against sadism and selfishness. See Rorty, "The Intellectuals at the End of Socialism," *Yale Review* 80 (1992): 1–16; Rorty, "Human Rights, Rationality, and Sentimentality," *Yale Review* 81 (1993): 1–20; and Rorty, "The Intellectuals and the Poor," a lecture delivered at Pomona College on February 19, 1996.

Contributors

ALLAN M. BRANDT is Amalie Moses Kass Professor of the History of Medicine at the Harvard University Medical School. He is the author of *No Magic Bullet: A Social History of Venereal Disease in the United States since 1880* (New York: Oxford University Press, 1985) and of articles in a wide range of publications on venereal disease, AIDS, and health risks, including "AIDS and Metaphor: Toward the Social Meaning of Epidemic Disease," *Social Research* 55 (1988): 413–32.

JOANNE BROWN is associate professor of history at the Johns Hopkins University. She is author of *The Definition of a Profession: The Authority of Metaphor in the History of Intelligence Testing, 1890–1930* (Princeton: Princeton University Press, 1992); she is coeditor of *The Estate of Social Knowledge* (Baltimore: Johns Hopkins University Press, 1991).

DAVID A. HOLLINGER is professor of history at the University of California, Berkeley. He is author of a number of studies in the intellectual history of nineteenth- and twentieth-century America and of several interventions in contemporary cultural and historiographical debates. His most recent book is *Postethnic America: Beyond Multiculturalism* (New York: Basic Books, 1995).

JAMES T. KLOPPENBERG is associate professor of history at Brandeis University. He is author of *Uncertain Victory: Social Democracy and Progressivism in European and American Thought, 1870–1920* (New York: Oxford University Press, 1986) and of articles in intellectual and cultural history, including "Objectivity and Historicism: A Century of American Historical Writing," *American Historical Review* 94 (1989): 1011–30.

ROLAND MARCHAND is professor of history at the University of California, Davis. He is author of *Advertising the American Dream: Making Way for Modernity, 1920–1940* (Berkeley: University of California Press, 1985) and other publications, including "The Corporation Nobody Knew: Bruce Barton, Alfred Sloan, and the Founding of the General Motors Family," *Business History Review* 65 (1991): 825–75.

REGINA MORANTZ-SANCHEZ is professor of history at the University of Michigan. She works primarily in the fields of the history of women, health, women physicians, and sexuality. Among her publications are *In Her Own Words: Oral Histories of Women Physicians* (Westport, Conn.: Greenwood, 1982) and *Sympathy and Science: Women Physicians in the American Medical Profession* (New York: Oxford University Press, 1985).

DOROTHY ROSS is Arthur O. Lovejoy Professor of History at the Johns Hopkins University. She is the author of *G. Stanley Hall: The Psychologist as Prophet* (Chicago: University of Chicago Press, 1972) and *The Origins of American Social Science* (New York: Cambridge University Press, 1991); she is the editor of *Modernist Impulses in the Human Sciences, 1870–1930* (Baltimore: Johns Hopkins University Press, 1993).

MICHAEL L. SMITH is associate professor of history at the University of California, Davis. He specializes in the cultural and political history of twentieth-century America, the history of technology, and the history of the American West. He is author of *Pacific Visions: California Scientists and the Environment, 1850–1915* (New Haven: Yale University Press, 1987).

JOHN R. STILGOE is a professor in the Department of Visual and Environmental Studies at Harvard University. He is author of numerous books and articles on the American landscape, including *Alongshore* (New Haven: Yale University Press, 1994), *Borderland: Origins of the American Suburb, 1820–1939* (New Haven: Yale University Press, 1988), and *Common Landscape of America, 1580 to 1845* (New Haven: Yale University Press, 1982).

RONALD G. WALTERS is professor of history at the Johns Hopkins University. Among his works are *American Reformers: 1815–1860* (New York: Hill and Wang, 1978), which is soon to appear in a revised edition, and *The Antislavery Appeal: American Abolitionism after 1830* (Baltimore: Johns Hopkins University Press, 1976). He is currently working on a study of twentieth-century American popular culture and is coediting, with Joan Cashin, a series on gender relations in the American experience for the Johns Hopkins University Press.

Index

Adams, Henry, 42, 43
Addams, Jane, 71
advertising, 51, 54, 56–81, 152, 155, 158, 161, 162, 163, 164, 169, 171. *See also* consumer culture; expositions; fairs; World's Fairs
affiliation, 19, 25, 30, 44
After Virtue (MacIntyre), 188
agricultural experiment stations, 121, 122, 137–38, 140–41, 142
agricultural press, 124–30
agricultural science, 5, 9, 118, 119–23, 137–42. *See also* farmers; U.S. Department of Agriculture
agriculture, 119–47
AIDS, 96
alcohol: consumption, 90, 93; dependency, 85
Alcoholics Anonymous, 85
Allen and Company, S. L., 132–37
American Association for the Advancement of Science, 152
American Dilemma (Myrdal), 13
American exceptionalism, 7, 33, 38–49
American Medical Association (AMA), 4, 76, 79, 80
American Medical Women's Association, 114
American School of Anthropology, 3
American Social Hygiene Association, 64
American Telegraph and Telephone (AT&T), 164, 168, 175
Anatomy of Antiliberalism (Holmes), 188
antibiotics, 86
antivivesection movement. *See* vivisection, opposition to
Atomic Energy Commission, 171
Austin, J. L., 192
Ayer and Son, N. W., 164

Bacon, Francis, 109
Bailey, Liberty Hyde, 138–39
Barton, Bruce, 152, 155, 161
Barton, Durstine, and Osborne, 161
Becker, Carl, 20
Bel Geddes, Norman, 164, 166, 167, 174
Bell Telephone Laboratories, 148
Bellah, Robert, 15, 197
Benedict, Ruth, 21
Bentley, Arthur F., 44–45
Bernstein, Richard, 189, 192, 197, 199
biomedical model. *See* germ theory; medicine, laboratory
biostatistics, 88. *See also* statistical methods
Blackwell, Elizabeth, 51, 52, 95–115; conception of maternity, 108, 109, 112
Blier, Ruth, 107
Blue Cross and Blue Shield, 97
Boas, Franz, 16, 21, 48
body, control over, 94, 95, 98
Borden corporation, 170
Bourne, Randolph, 28
Boyd, T. A., 159
Brandeis, Louis, 191
Brandt, Allan M., 4, 51, 52, 73, 261
Brinkley, David, 169
Bristol-Meyers, 78
Brown, JoAnne, 4, 7, 51, 52, 261
Bryant, Christopher G. A., 46
Burnham, John C., 118, 149, 159, 163
Bush, Vannevar, 7

Califano, Joseph, 91
Cameron, W. J., 166
Campbell, Joseph, 26
cancer, 84, 86, 88, 91, 92

The Library of Congress has cataloged the hardcover edition of this
book as follows:

Scientific authority and twentieth-century America / edited by Ronald
   G. Walters.
      p.   cm.
   Includes bibliographical references and index.
   ISBN 0-8018-5389-3 (alk. paper).
     1. Science—Social aspects—United States—History—20th century.
2. Medicine—Social aspects—United States—History—20th century.
3. United States—Intellectual life—20th century.   I. Walters,
Ronald G.
Q175.52.U5S38   1997
303.48´3´0973—dc21   96-46908

ISBN 0-8018-5390-7 (pbk.)